202
Topics in Current Chemistry

Springer

Berlin
Heidelberg
New York
Barcelona
Hong Kong
London
Milan
Paris
Singapore
Tokyo

Implementation and Redesign of Catalytic Function in Biopolymers

Volume Editor: F. P. Schmidtchen

With contributions by
L. Baltzer, A. R. Chamberlin, K. A. McDonnell,
M. Famulok, M. A. Gilmore, B. Imperiali, A. Jenne,
M. Shogren-Knaak, L. E. Steward

 Springer

This series presents critical reviews of the present position and future trends in modern chemical research. It is addressed to all research and industrial chemists who wish to keep abreast of advances in the topics covered.

As a rule, contributions are specially commissioned. The editors and publishers will, however, always be pleased to receive suggestions and supplementary information. Papers are accepted for "Topics in Current Chemistry" in English.

In references Topics in Current Chemistry is abbreviated Top. Curr. Chem. and is cited as a journal.

Springer WWW home page: http://www.springer.de
Visit the TCC home page at http://link.springer.de/series/tcc

ISSN 0340-1022
ISBN 3-540-65728-2
Springer-Verlag Berlin Heidelberg New York

Library of Congress Catalog Card Number 74-644622

Cover design: Friedhelm Steinen-Broo, Barcelona; MEDIO, Berlin
Typesetting: Fotosatz-Service Köhler GmbH, 97084 Würzburg

SPIN: 10649327   02/3020 – 5 4 3 2 1 0 – Printed on acid-free paper

Topics in Current Chemistry
Now Also Available Electronically

For all customers with a standing order for Topics in Current Chemistry we offer the electronic form via LINK free of charge. Please contact your librarian who can receive a password for free access to the full articles. By registration at:

http://link.springer.de/series/tcc/reg_form.htm

If you do not have a standing order you can nevertheless browse through the table of contents of the volumes and the abstracts of each article at:

http://link.springer.de/series/tcc

There you will also find information about the

- Editorial Board
- Aims and Scope
- Instructions for Authors

Preface

Even today low molecular weight enzyme models generally still do not have the power of natural biocatalysts. While certain aspects of biocatalysis like substrate-, chemo-, regio- or stereoselectivity or even regulatory features can be mimicked at least phenomenologically by abiotic counterparts, it is the absolute magnitude of the catalytic activity that eludes imitation by artificial enzyme surrogates. Thus, it seems debatable whether or not a simple reductionistic approach to understanding biocatalysis must be supplemented or even replaced by concepts that include and consider the polymeric nature of biocatalysts and the properties derived therefrom.

Following this reasoning, a rational route for proceeding calls for the deliberate and prudent exchange of functions or structural motifs or the addition of new ones in fully functional biopolymers and observe the consequences in terms of stability and catalytic activity. There is hope that a limited structural modification at one particular site will entail a locally limited response that can be dissected and analyzed. The results emerging in the context of the functional catalyst are expected to be more readily translated into measures to be taken for the improvement of catalytic function.

Prerequisites to this approach are methods of incorporating non-natural moieties at predetermined positions in the biopolymer. This volume is meant to serve as a source in this respect describing the state of the art of some major lines of attack with this goal in mind. As a modern alternative, the creation of novel catalysts by directed evolution of nucleic acid aptamers is included. In this case, too, it is of prime importance to learn about the structural details which cooperate to bring about the catalytic function underlying the selection process.

Garching, April 1999   Franz P. Schmidtchen

Contents

Contents of Volume 195

Biosynthesis
Polyketides and Vitamins

Contents of Volume 200

Biocatalysis
From Discovery to Application

Design and Construction of Novel Peptides and Proteins by Tailored Incorporation of Coenzyme Functionality

Barbara Imperiali* · Kevin A. McDonnell · Michael Shogren-Knaak

Department of Chemistry, Massachusetts Institute of Technology, Cambridge, MA 02139, USA.
E-mail: imper@mit.edu

Recent research in a number of laboratories has focused on devising strategies for the integration of coenzyme functionality into non-native biopolymer scaffolds. The goal of this research is to influence the inherent chemical reactivity of coenzymes to generate unique molecules with novel functional properties. This review will discuss the progress made in this area and delineate the issues involved in using polypeptides and proteins to harness the chemical properties of selected coenzymes. The specific coenzymes that will be presented in this review are pyridoxal phosphate, thiazolium pyrophosphate, flavin and nicotinamide adenine dinucleotide.

Keywords: Protein design, Coenzyme-dependent enzyme, Pyridoxal, Pyridoxamine, Thiamine pyrophosphate, Nicotinamide adenine dinucleotide, Flavin, Semisynthesis, Polypeptide motif.

---

* Corresponding author.

Topics in Current Chemistry, Vol. 202
© Springer-Verlag Berlin Heidelberg 1999

List of Abbreviations

CD	circular dichroism
FAD	flavin adenine dinucleotide
FMN	flavin mononucleotide
NAD	nicotinamide adenine dinucleotide
NADP	nicotinamide adenine dinucleotide phosphate
NMR	nuclear magnetic resonance
dPam	deoxypyridoxamine
TDP	thiamine diphosphate
ee	enantiomeric excess
SPPS	solid phase peptide synthesis
Glc	*N*-terminal glycolate capping group
Dap	α,β-diaminopropionic acid
Dab	α,γ-diaminobutyric acid
Hcy	homocysteine
Rnase S	ribonuclease S
IFABP	intestinal fatty acid binding protein
ALBP	adipocyte lipid binding protein
NOE	nuclear Overhauser effect
NAH	dihydronicotinamide
UV/vis	ultraviolet and visible
DNA	deoxyribonucleic acid
EDTA	ethylenediaminetetraacetic acid
HPLC	high performance liquid chromatography
SDS	sodium dodecyl sulfate
Boc	*tert*-butyloxycarbonyl
Fmoc	*N*-fluorenylmethoxycarbonyl
Ac	acetyl
TFA	trifluoroacetic acid
DIPEA	*N,N,N*-diisopropylethylamine
PNP	*p*-nitrophenol
Dopa	dopamine

1
Introduction

Coenzymes are densely functionalized organic cofactors capable of catalyzing numerous diverse chemical reactions. Nature exploits the intrinsic chemical reactivity of these molecules to extend the chemical functionality of enzymes well beyond the reactivity of the coded amino acids. When these constituents are incorporated via covalent or non-covalent interactions into coenzyme-dependent enzymes, the inherent reactivity of the coenzyme is augmented and directed to effect chemical transformations with substrate and product selectivities, rates, and yields that are unachievable by either the protein or coenzyme alone. Thus, coenzymes play a critical role in the execution of a large number of essential metabolic processes.

Although the chemical reactivity of many coenzymes could be exploited in the design of new functional proteins, the use of coenzymes has been limited by the need for explicit binding sites in the host protein. However, research in a number of laboratories has focused on developing new strategies for the integration of coenzyme functionality into biopolymer scaffolds with the goal of modulating the inherent chemical reactivity of coenzymes to generate unique molecules with novel functions. This review will present some of the progress made in this area and delineate the issues involved in using polypeptides and proteins to harness the chemical properties of selected coenzymes.

Coenzymes facilitate chemical reactions through a range of different reaction mechanisms, some of which will be discussed in detail in this review. However, in all cases structural features of the coenzyme allow particular reactions to proceed along a mechanistic pathway in which reaction intermediates are more thermodynamically and kinetically accessible. When incorporated into apoenzyme active sites, the coenzyme reactivity is influenced by a well-defined array of amino acid functional groups. For a given coenzyme, the particular array of amino acids presented by the different apoenzymes can drastically alter the degree of rate acceleration and product turnover and can specify the nature of the reaction catalyzed.

While nature uses coenzyme-dependent enzymes to influence the inherent reactivity of the coenzyme, in principle, any chemical microenvironment could modulate the chemical properties of coenzymes to achieve novel functional properties. In some cases even simple changes in solvent, pH, and ionic strength can alter the coenzyme reactivity. Early attempts to present coenzymes with a more complex chemical environment focused on incorporating coenzymes into small molecule scaffolds or synthetic host molecules such as cyclophanes and cyclodextrins [1, 2]. While some notable successes have been reported, these strategies have been less successful for constructing more complex coenzyme microenvironments and have suffered from difficulties in readily manipulating the structure of the coenzyme microenvironment.

Peptide and protein scaffolds offer a means to influence coenzyme reactivity without some of the limitations of the host organic scaffolds described above. Both proteins and peptides can be used to generate complex chemical environments. Existing proteins, which already contain structural complexity, can be altered or redesigned to accommodate a coenzyme functionality. Alternately, recent progress in creating structured polypeptide motifs [3–6] indicates that smaller non-native systems could be used to present a complex microenvironment to the coenzyme functionality. Protein and peptide scaffolds are advantageous because they offer a facile means of changing the chemical environment of a coenzyme; the modular nature of proteins and peptides allows the introduction of a large range of standard, modified, and unnatural amino acids. Moreover, these changes can be rapidly effected using well-developed techniques of chemical synthesis, biochemistry, and molecular biology. Finally, proteins and peptides are attractive constructs for influencing coenzyme reactivity because the structural features of these molecules can be elucidated using standard biophysical methods, including NMR spectroscopy and X-ray crystallo-

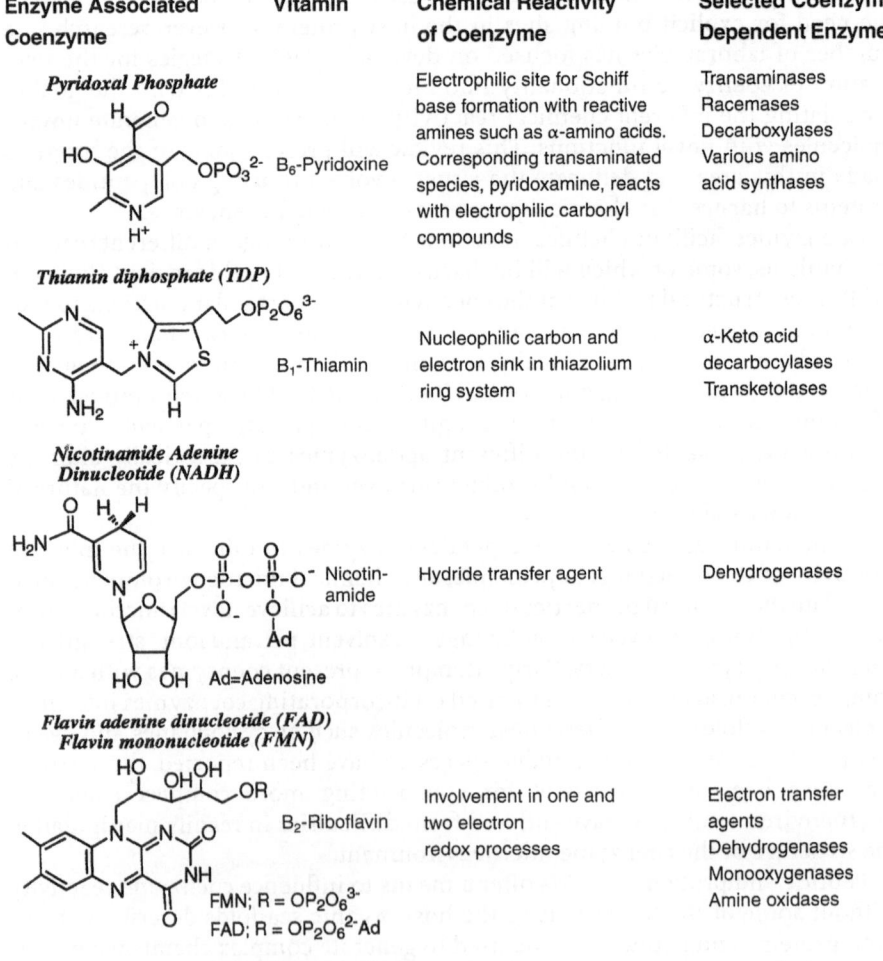

Enzyme Associated Coenzyme	Vitamin	Chemical Reactivity of Coenzyme	Selected Coenzyme Dependent Enzymes
Pyridoxal Phosphate	B_6-Pyridoxine	Electrophilic site for Schiff base formation with reactive amines such as α-amino acids. Corresponding transaminated species, pyridoxamine, reacts with electrophilic carbonyl compounds	Transaminases Racemases Decarboxylases Various amino acid synthases
Thiamin diphosphate (TDP)	B_1-Thiamin	Nucleophilic carbon and electron sink in thiazolium ring system	α-Keto acid decarbocylases Transketolases
Nicotinamide Adenine Dinucleotide (NADH)	Nicotinamide	Hydride transfer agent	Dehydrogenases
Flavin adenine dinucleotide (FAD) Flavin mononucleotide (FMN)	B_2-Riboflavin	Involvement in one and two electron redox processes	Electron transfer agents Dehydrogenases Monooxygenases Amine oxidases

Fig. 1. Summary of coenzymes discussed in this review

graphy. These studies can offer critical structure-function information to better understand how changes in a coenzyme environment create functional properties. This provides insight into the design of new molecules with improved function.

This review will explore some of the issues involved in developing novel peptides and proteins that integrate coenzyme functionality and highlights recent progress in this area. For the sake of depth, this review focuses specifically on the pyridoxal/pyridoxamine system, thiamine, and selected redox active coenzymes. These coenzymes and information on the transformations that they mediate are summarized in Fig. 1.

2
Pyridoxal and Pyridoxamine

2.1
Pyridoxal and Pyridoxamine in Solution and in Enzymes

The pyridoxal and pyridoxamine coenzymes are critical for the enzymatic processing of α-amino acids. Pyridoxal facilitates production of a range of products by effectively stabilizing a carbanion intermediate at the α-carbon of an α-amino acid. If this carbanion equivalent is generated by loss of the α-carboxylate of the amino acid, reprotonation of the carbanion equivalent results in the amino acid decarboxylation product. When the carbanion equivalent results from loss of the α-proton, a number of different reaction products can result. Reprotonation of this site results in racemization or epimerization. Elimination of a leaving group at the amino acid β-carbon generates the enamine equivalent, which can accept a nucleophile to generate the β-replacement product or tautomerize and hydrolyze to the β-elimination product. Finally, a more complicated sequence of mechanistic steps results in the conversion of an α-amino acid to the corresponding α-keto acid by transamination.

Transamination is the most thoroughly studied pyridoxal-mediated process; an examination of the reaction illustrates how the cofactor, together with the cognate apoenzyme, both stabilizes and selects intermediates along the reaction pathway. Moreover, transamination is the principle reaction that has been studied in the context of design and redesign of pyridoxal-containing proteins and peptides. In a typical transamination reaction, one α-amino acid (amino acid$_1$) is converted into an α-keto acid (keto acid$_1$). A different α-keto acid (keto acid$_2$) is subsequently converted into a new α-amino acid (amino acid$_2$) as illustrated in Fig. 2. In this process pyridoxal acts as a catalytic intermediate, being converted from the aldehydic form to the reduced pyridoxamine form and then back again.

Mechanistically, transamination by the free coenzyme proceeds through a number of discrete steps as illustrated in Fig. 3. The first step of the process, aldimine formation (Fig. 3, Step I), is common to all pyridoxal-dependent reactions. The rate and extent of this reaction are influenced by factors including reactant concentrations, the nature of the amino acid, pH, and solvent. However, it is important to realize that the coenzyme itself facilitates Schiff base formation in

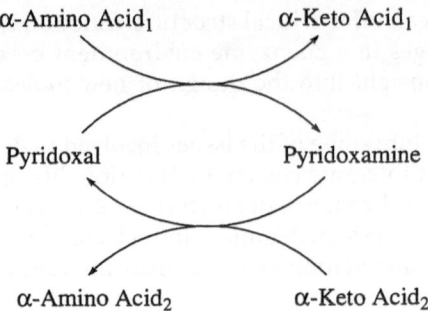

Fig. 2. Catalytic cycle of pyridoxal/pyridoxamine-dependent transamination

Pyridoxal   Amino Acid   Aldimine   Quinoid

Pyridoxamine   Keto acid   Ketimine

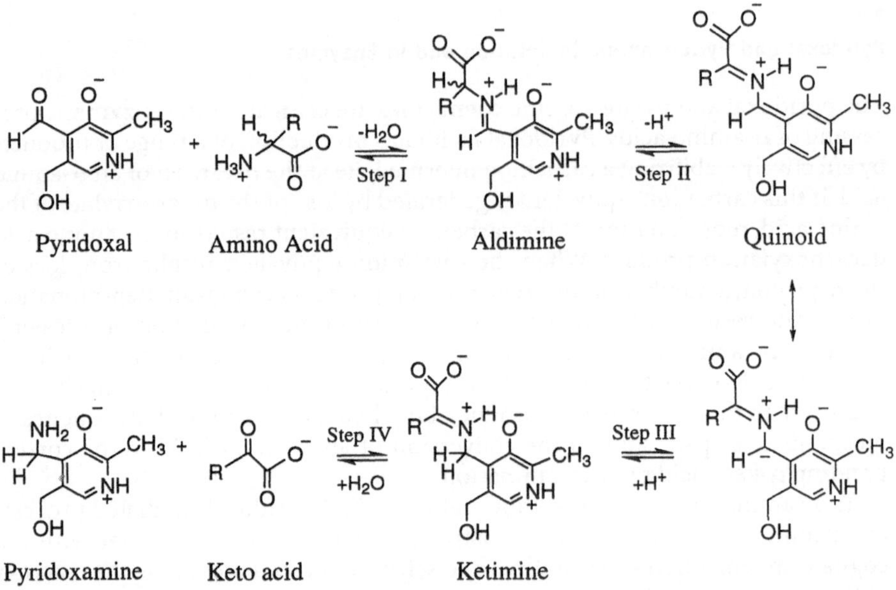

Fig. 3. Reaction pathway and intermediates in pyridoxal/pyridoxamine-dependent transamination

a number of ways, including activation of the pyridoxal aldehyde and stabilization of the aldimine intermediate by promoting protonation of the Schiff base nitrogen [7]. Once the aldimine is formed, deprotonation (Fig. 3, Step II) is promoted by electronic delocalization across the ring system [8]. This delocalization is further facilitated through protonation of the pyridyl nitrogen of the aldimine, thereby allowing the ring to act as an "electron sink" to form the quinoid intermediate [7]. The quinoid intermediate is selectively reprotonated at the 4'-position (Fig. 3, Step III) to afford the ketimine intermediate. Subsequent hydrolysis (Fig. 3, Step IV) generates pyridoxamine and the new α-amino acid.

To achieve rapid reaction rates in addition to reactant and product selectivity, the structure of the apoenzyme is thought to both augment and limit the mechanistic pathways available to the coenzyme [9]. Both the substrate and coenzyme are tightly held by the enzyme. Pyridoxal phosphate is bound to the protein through a number of interactions, primarily between the guanidinium group of an arginine and the phosphate of the coenzyme [10]. Additionally, the coenzyme is covalently tethered to a lysine side chain through an internal Schiff base linkage. By directly tethering the coenzyme functionality to the protein platform, many of the transaminase model systems described in this chapter circumvent the necessity for discrete coenzyme binding interactions. The α-amino acid substrate is bound through interactions between the α-carboxylate of the amino acid and the guanidinium group of an arginine side chain, as well as through other interactions [10]. By binding both the α-amino acid substrate and pyridoxal phosphate, the rate and stability of aldimine formation (Fig. 3, Step I) are augmented by an increase in effective concentration and by promoting a reactive orientation. These features can potentially be emulated in designed systems.

After formation of the aldimine, numerous factors in the enzyme facilitate deprotonation of the α-carbon (Fig. 3, Step II). The lysine liberated by transimination is utilized as a general base and is properly oriented for effective deprotonation [11]. Furthermore, the inductive effects of the ring system are tuned to increase the stabilization of the quinoid intermediate. For example, the aspartate group that interacts with the pyridyl nitrogen of the coenzyme promotes protonation to allow the ring to act as a more effective electron sink. In contrast, in alanine racemase, a less basic arginine residue in place of the aspartic acid is believed to favor racemization over transamination [12].

Product formation by the transaminase is accomplished by using the newly protonated lysine group as a general acid (Fig. 3, Step III). This residue is well positioned to act as a proton shuttle, moving the proton from the former α-carbon position to the 4'-position of the ring. When this series of reactions is run in reverse to generate an amino acid from a keto acid, the facial selectivity of the process allows only one enantiomeric form of the product to be generated [10]. Additionally, the coenzyme active site excludes water, preventing the solvent from reprotonating the quinoid intermediate non-selectively [10].

Powerful mechanistic enzymology in conjunction with structural biology studies have provided insight into the manner in which the protein environment of an apoenzyme augments and controls coenzyme reactivity. The goal of the protein design efforts that include coenzyme functionality is to establish whether these features can be emulated within designed peptide and protein constructs.

2.2
Pyridoxal Associated Peptide Constructs

Baltzer and co-workers have utilized a synthetic peptide scaffold to incorporate features of transaminase enzymes [13]. Using this approach, they have attempted to achieve the selective and tight binding of a pyridoxal phosphate coenzyme observed in transaminase enzymes. A synthetic helix-turn-helix peptide known to dimerize into a four-helix bundle was chosen as the platform for design [5].

Three lysine groups were incorporated at positions 11, 15, and 30 on the solvent-exposed face of the helices (Fig. 4). These residues were chosen to form an internal aldimine similar to that observed in the active site lysine of native transaminases. To generate selective and strong binding to the lysine located at position 30, an arginine group was incorporated into position 19. Molecular modeling revealed that in the folded motif, the guanidinium group of this residue would be expected to form electrostatic interactions with the phosphate group of the coenzyme. Spectroscopic analysis of the synthetic peptide, PP-42, indicated that these changes could be made to the synthetic template without loss of helicity or loss of association between the monomer units.

Peptide PP-42 demonstrated enhanced binding of pyridoxal phosphate. Both NMR and UV/vis spectroscopic studies indicated that, under acidic reaction conditions, more than 95 % of the peptide is in the aldimine form. Moreover, when similar conditions were used with pyridoxal, which lacks the phosphate, no aldimine complex was formed. While this difference in binding affinity could not be quantified, it constituted at least a two-order of magnitude difference in the equilibrium binding constants. Thus, the phosphate group appears to generate a significant increase in affinity of the coenzyme for the peptide and supports the proposal that the guanidinium group of arginine 19 provides the driving force for this effect.

Binding of pyridoxal phosphate to peptide PP-42 also appears to be selective for lysine 30. As was indicated by NMR spectroscopy and UV/vis experiments, only one of three potential lysine Schiff bases appeared to form. To determine the site or sites of attachment, the aldimine peptide intermediates were reduced, proteolytically cleaved, and the fragments analyzed by mass spectroscopy. This

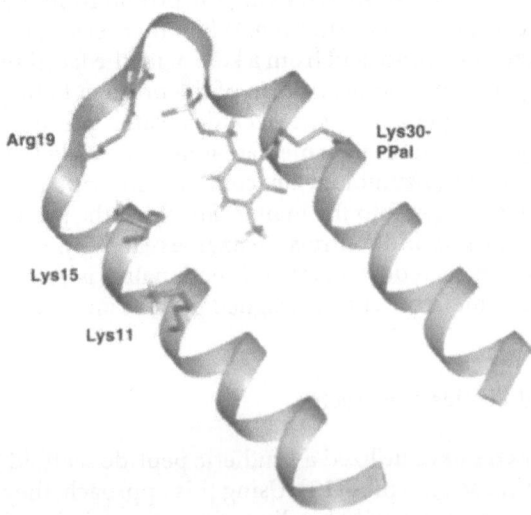

Fig. 4. Proposed structure of PP-42 with depiction of the relevant lysine and arginine residues. Reproduced with permission from J Chem Soc Chem Commun (1998) 15:1547

analysis indicated that approximately 85% of the pyridoxal phosphate was associated with lysine 30, 15% with lysine 11, and virtually none with lysine 15.

This designed construct, as such, may not influence subsequent steps of transamination due to loss of the strong association between the coenzyme and the synthetic peptide after amino acid binding. However, the selectivity of the peptide for pyridoxal phosphate reveals the potential power of peptide design and the importance of secondary binding interactions for defining the formation of specific binary complexes.

2.3
Tethered Pyridoxamine Proteins

Distefano and co-workers have concentrated on using a native protein scaffold to influence aspects of pyridoxamine-mediated transamination [14–18]. Specifically, these researchers sought to increase rates of product formation, stereoselectivity, and multiple product turnovers. To achieve these goals, they introduced the coenzyme functionality into a protein by chemical modification. As platforms for the coenzyme, the group utilized adipocyte lipid binding protein (ALBP) and intestinal fatty acid binding protein (IFABP). These related lipid binding proteins are attractive protein scaffolds because they have been structurally well characterized by X-ray crystallography [19]. Additionally, both ALBP and IFABP are small 15 kDa proteins which can be produced by efficient expression systems. The structure of these proteins is comprised of two perpendicular five-stranded β-sheet regions that enclose a 300–500 Å^3 hydrophobic cavity (Fig. 5). This cavity is large enough to encapsulate a wide range of fatty

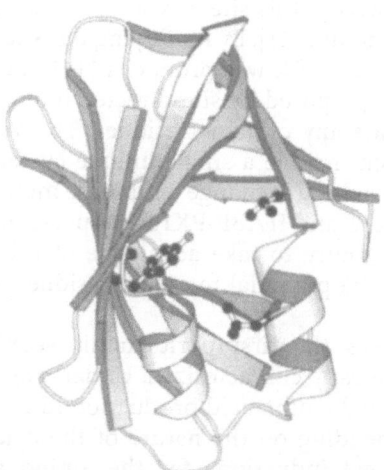

Fig. 5. Stereo view of IFABP. Ball-and-stick residues depict the sites of cysteine incorporation and modification. Clockwise from top right: residues Val60, Leu72, Tyr117 and Ala104. Residue Tyr117 has been included to show where Cys117 would be located in the structurally similar protein ALBP. Reproduced with permission from Bioorg Med Chem Lett (1997) 7:2055

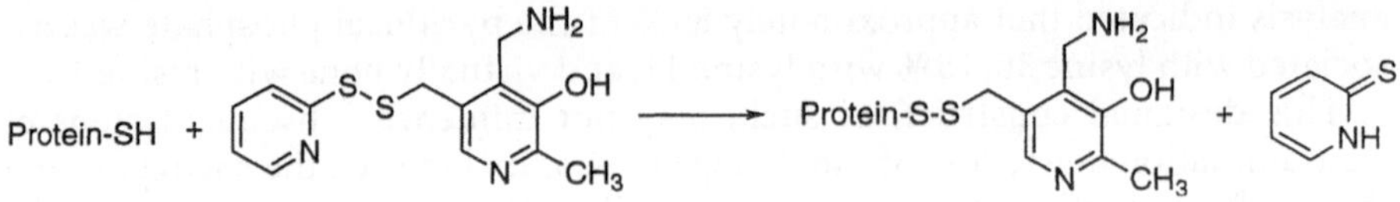

Fig. 6. Reaction for the preparation of pyridoxamine-protein conjugates

acid substrates and can potentially serve as a enclosed protein microenvironment for a pyridoxal/pyridoxamine cofactor. Site-directed mutagenesis and sequence homology studies have also established which residues within the protein cavity can be changed without loss of structure. Consequently, appropriate sites for tethering the coenzyme functionality could be readily identified.

Incorporation of the coenzyme functionality into ALBP and IFABP was accomplished by chemical modification of a cysteine residue with a reactive pyridoxamine derivative in a manner analogous to that developed by Kaiser and co-workers as illustrated in Fig. 6 [20]. This pyridoxamine derivative was synthesized from pyridoxamine and includes a redox activated 2-pyridyl-dithiol group in place of the 5'-hydroxyl group of the original cofactor [14]. Disulfide exchange with a protein-bound cysteine group covalently tethers the modified pyridoxamine group to the protein via a disulfide linkage. A high degree of derivatization was confirmed by both spectroscopic and chemical techniques.

Sites for coenzyme incorporation were chosen to minimize the structural perturbations in the protein and to generate potentially interesting function. In the native form, ALBP contains a single cysteine residue at position 117 in the interior of the protein cavity; this site was shown to accept disulfide modifications without loss of structure [21]. By modifying this site with the pyridoxamine derivative, the protein ALBP-PX was produced [14]. Additionally, a number of mutants of IFABP were prepared by site-directed mutagenesis since the native protein does not include any cysteine residues. The Val60Cys, Leu72Cys, and Ala104Cys mutants each include a single unique cysteine in the hydrophobic cavity of the protein. Disulfide exchange with these mutants resulted in proteins IFABP-PX60, IFABP-PX72, and IFABP-PX104, respectively [15]. All of these sites were chosen for their ability to take advantage of the asymmetric nature of the protein cavity and the potential for some residues to interact with the co-enzyme.

Protein ALBP-PX was the first pyridoxamine-conjugated protein to be synthesized and structurally characterized. Under single-turnover conditions, this protein demonstrated amino acid production rates of only 56% of the free cofactor. However, depending on the nature of the α-keto acid, ALBP-PX did show a range of optical inductions for the amino acid product. Notably, enantiomeric excesses in the order of 94% were observed for the production of valine. Additionally, several trends were noted. All amino acid products that showed optical induction favored the l-enantiomer, except alanine, which favored the d-enantiomer. Furthermore, α-keto acids with branched side chains showed greater optical induction, with the β-branched precursor of valine showing the highest optical induction.

From the crystal structure of ALBP-PX, a number of observations were made [17]. As designed, the modified pyridoxamine functionality is contained within the hydrophobic interior of the ALBP protein. This substitution results in only minor perturbations of the overall protein structure relative to the native protein. In the cavity, the modified cofactor is surrounded by a number of hydrophobic residues. Additionally, the coenzyme derivative has some interactions with more polar residues; for example, the exocyclic amine of pyridoxamine appears to be hydrogen bonded to both arginine 126 and tyrosine 128. Interestingly, hydrogen bonding from these residues holds the exocyclic amine in an orientation that is not planar with the pyridyl ring system. Thus, this amine orientation is probably not the geometry that is present during deprotonation of the 4'-methylene, which is promoted by a planar ketimine intermediate.

To explain the observed optical induction, a substrate was incorporated into the molecular model of the protein. A substrate such as α-ketoglutarate could be included in the protein model with a geometry that allowed stereoselective protonation of the quinoid intermediate by solvent, consistent with the enantiomeric excess (ee) of the l-stereoisomer product. Moreover, the geometry consistent with production of the d-enantiomer appeared too sterically crowded for most substrates. However, pyruvic acid, which was the only substrate to favor the d-enantiomer product, was small enough to adopt the alternative geometry and also had the potential to interact with an arginine group.

Characterization of the IFABP-derivatized proteins showed promising results. While IFABP-PX104 exhibited little reactivity and IFABP-PX73 showed activity comparable to ALBP-PX, IFABP-PX60 afforded nearly quantitative yields of glutamic acid from α-ketoglutarate under single-turnover conditions during the assay period [15]. Transamination of α-ketoglutarate by IFABP-PX60 was investigated further under turnover conditions in the presence of a variety of added hydrophobic amino acids [16]. Optical inductions of at least 70% for l-glutamic acid were observed for all of the amino acids used. The results from these experiments are summarized in Table 1.

Table 1. IFABP-PX60-mediated transamination of α-ketoglutarate to glutamic acid in the presence of various hydrophobic amino acids

Amino acid	Turnover(s) (n)	ee (%)
none	> 0.99	68 (L)
Ala	1.30 ± 0.05	70 (L)
Val	1.1 ± 0.03	79 (L)
Leu	1.6 ± 0.11	80 (L)
Phe	3.9 ± 0.15	93 (L)
Tyr	4.2 ± 0.07	93 (L)
Trp	2.2 ± 0.04	92 (L)
Dopa	4.3 ± 0.12	94 (L)

Aromatic amino acids afforded the greatest optical induction, with tyrosine and phenylalanine producing a 94% ee of l-glutamic acid. Use of tyrosine generated the greatest yield of the glutamate product, giving several turnovers during the assay. The nature of the observed rate acceleration was determined by performing kinetic analysis with varying substrate concentrations and fitting the results to a Michaelis–Menten model. From this analysis, it was determined that K_m for the protein decreased by 52-fold, while k_{cat} increased by 3.9-fold. Thus, the observed rate acceleration from IFABP-PX60 can be interpreted as partially resulting from an increased rate of transamination, and largely from an increase in binding affinity for substrate.

2.4
Pyridoxal and Pyridoxamine Amino Acid Chimera Containing Peptides

Work in the Imperiali laboratory has also focused on exploring the ability of minimal peptide scaffolds to augment the rate of coenzyme-mediated transaminations [22–25]. To accomplish this, a strategy has been developed in which the core functionality of the coenzyme is incorporated as an integral constituent of an unnatural coenzyme amino acid chimera construct. Thus, non-covalent binding of the coenzyme to the peptide or protein scaffold is unnecessary. Both the pyridoxal and pyridoxamine analogs have been synthesized in a form competent for Fmoc-based solid phase peptide synthesis (SPPS) (Fig. 7) [23,24].

The pyridoxal amino acid analog (Pal) was stereoselectively synthesized from a readily available pyridoxol derivative and the residue was incorporated into peptides at the alcohol oxidation state in protected form. Oxidation of the 4′-alcohol group to the desired aldehyde was achieved post-synthetically on free,

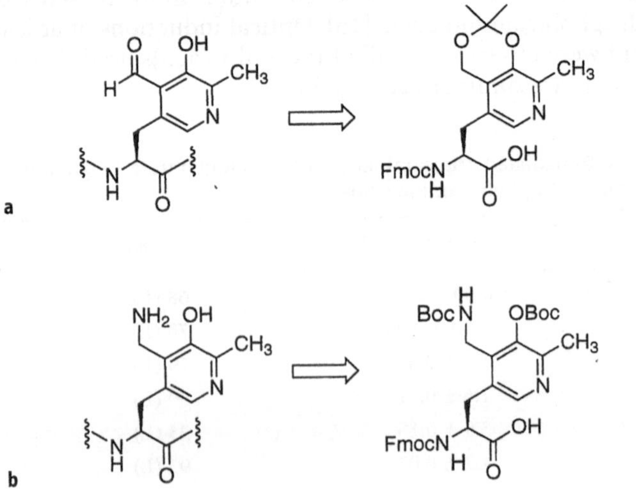

Fig. 7. Amino acid coenzyme chimeras of *a* pyridoxal and *b* pyridoxamine in both peptide incorporated and SPPS competent form

deprotected peptide using γ-MnO$_2$. The pyridoxamine amino acid analog (Pam) was also synthesized stereospecifically. The advantage of Pam is that it does not require a post-synthesis oxidation for conversion to the chemically active form of the coenzyme derivative.

One advantage of the coenzyme amino acid chimera approach is that it is compatible with solid phase peptide synthesis. Consequently, the reactive functionality can be readily and selectively delivered to any site in the peptide. Additionally, both natural and unnatural residues can be incorporated throughout the peptide scaffold, and related compounds can be investigated rapidly by combinatorial synthesis techniques.

The simplest structural platforms to be exploited with the coenzyme amino acid chimeras were a family of β-hairpin peptides. In these peptides, the amino acid side chains are oriented perpendicular to the plane created by the hairpin, with sequential residue side chains positioned on alternating sides of the hairpin. By placing specific residues across the sheet from the coenzyme residue, these peptides offer a potentially simple way of delivering side-chain functionality to the coenzyme derivative. A β-hairpin peptide structure is promoted by the inclusion of a hetero-chiral turn sequence to promote type II or II' β-turn formation [26].

Early studies explored the ability of a general-base residue to augment the rate of transamination. Three hexapeptides, P1–P3 (illustrated in Fig. 8), with Pal at position 5 and a type II β-turn at positions 3 and 4, were constructed [23]. These peptides incorporate valine, histidine, or 3-(3-pyridyl)alanine across the sheet from Pal at position 2. Transamination of alanine to pyruvic acid suggests that the rate of transamination can be augmented to a modest degree by inclusion of a basic residue (Table 2). The 3-(3-pyridyl)alanine-containing peptide P3 shows a 4.7-fold rate enhancement relative to the model compound 5'-deoxypyridoxal (dPal). Poorer rate enhancements are observed for the histidine- and valine-containing peptides. Thus, these results correlate with the ability of the side chain to accept a proton under these reaction conditions. Furthermore, structure appears to be an important feature of the rate enhancement. When the type II turn residues are both replaced with glycine, the rate observed is comparable to that of the model compound alone.

The ability of a more extended coenzyme environment to influence coenzyme-mediated transamination was also explored using a semisynthetic protein con-

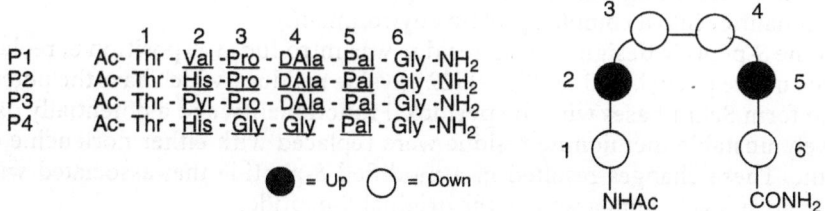

P1 Ac- Thr - Val -Pro - DAla - Pal - Gly -NH$_2$
P2 Ac- Thr - His -Pro - DAla - Pal - Gly -NH$_2$
P3 Ac- Thr - Pyr -Pro - DAla - Pal - Gly -NH$_2$
P4 Ac- Thr - His - Gly - Gly - Pal - Gly -NH$_2$

● = Up   ○ = Down

Fig. 8. Sequence of Pal-containing β-hairpin peptides P1–P4 and proposed patterning of side-chain residues

Table 2. Observed rate constants for the transamination of pyruvate to alanine mediated by various pyridoxal derivatives

Coenzyme System	$k_{obs} \cdot 10^5$ (min^{-1})
P1	165
P2	225
P3	278
P4	62
dPal	59

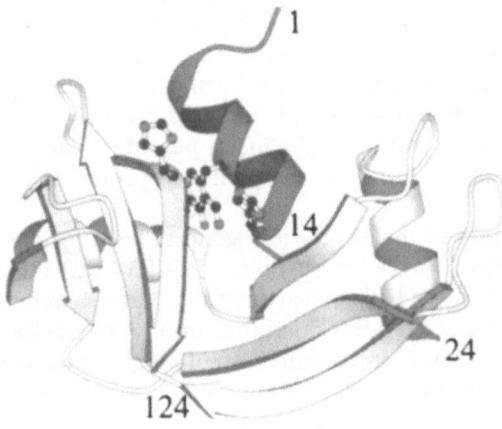

Fig. 9. Proposed structure for the semisynthetic RNase-S complex incorporating the Pal residue. S-peptide shown in *dark gray*. Reproduced with permission from J Am Chem Soc (1994) 116:12084

struct. The enzyme ribonuclease A can be subjected to limited proteolysis by trypsin to produce two fragments, a helical domain containing residues 1–20 (S-peptide) and a globular domain containing residues 21–124 (S-protein). These fragments can be reconstituted in a non-covalent complex to afford a fully active ribonuclease, RNase-S complex [27]. Thus, by synthesizing new S-peptides containing Pal or Pam, it is possible to introduce these coenzyme functionalities into a complex protein environment.

In the S-peptide design, the Pal residue was introduced at position 8, replacing the native phenylalanine (Fig. 9) [22]. Lysine residues, which have the potential to form Schiff bases with the pyridoxal functionality, and a potentially oxidatively unstable methionine residue were replaced with either norleucine or glycine. These changes resulted in a modified S-peptide that associated with S-protein at levels comparable to the original S-peptide.

The rate of alanine transamination to pyruvic acid was studied under single-turnover conditions and an 18-fold increase in rate was observed in the presence of S-protein. This data indicates that the microenvironment generated by the protein complex augmented the properties of the coenzyme functionality. Interestingly, replacing the lysine residue that originally flanked the coenzyme moiety with an isosteric norleucine residue resulted in an observed rate constant that was not perturbed by the S-protein, potentially due to steric occlusion. Incorporation of the Pam amino acid into the S-peptide resulted in complexes with the S-protein which demonstrated up to a 7.4-fold rate enhancement relative to the S-peptide alone.

More recently, the Pam amino acid chimera has also been incorporated into a small $\beta\beta\alpha$-motif peptide scaffold [28]. The family of BBA peptides was developed in our laboratory as structured platforms for the design of functional motifs. These motifs are attractive because they are small enough (23 residues) to be easily synthesized by standard solid phase synthesis methods. Additionally, the motifs appear to possess sufficient structural complexity to influence coenzyme properties while still being amenable to structural characterization by standard spectroscopic techniques [3, 29, 30]. The BBA peptides include a β-hairpin domain with a type II' turn connected by a loop region to an α-helical domain (Fig. 10). Packing of the sheet and helix against one another is accomplished by hydrophobic contacts created by a hydrophobic core of residues.

The Pam chimera was incorporated into several different positions in the BBA peptide sequence, including the middle of the α-helix domain at position 17 and in the hydrophobic core at position 8. However, the conservative replacement of tyrosine for Pam at position 6, the interface between the type II' turn

Glc-Tyr-Dap-Val-DPro-5-Pam-7-Phe-Ser-Arg-Ser-Asp-Glu-Leu-Ala-Lys-Leu-Leu-Arg-Leu-His-Ala-Gly-NH$_2$

a

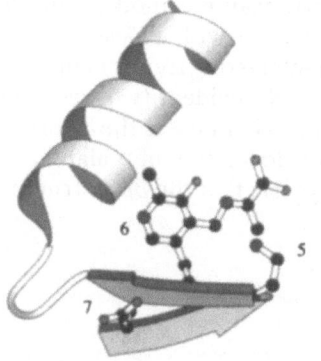

b

Peptide	Pos. 5	Pos. 7
CBP01	His	Asp
CBP02	Lys	Asp
CBP03	Orn	Asp
CBP04	Dab	Asp
CBP05	Cys	Asp
CBP06	Hcy	Asp
CBP07	His	Gly
CBP08	Lys	Gly
CBP09	Orn	Gly
CBP10	Dab	Gly
CBP11	Cys	Gly
CBP12	Hcy	Gly
CBP13	His	Dab
CBP14	Lys	Dab
CBP15	Orn	Dab
CBP16	Dab	Dab
CBP17	Cys	Dab
CBP18	Hcy	Dab

c

Fig. 10. *a* General sequence of peptides CBP01–CBP18. *b* proposed structure of peptide CBP04 (BP1) based on the structure of the parent peptide. Residues 5 through 7 are depicted in ball-and-stick representation, with Pam shown in the ketimine form. *c* table showing specific substitution at positions 5 and 7. *Glc* N-terminal capping group; *Dap* α,β-diaminopropionic acid; *Dab* α,γ-diaminobutyric acid; *Hcy* homocysteine

and the β-sheet domain, resulted in a peptide, BP1, that showed signs of both secondary and supersecondary $\beta\beta\alpha$-motif structure as assessed by circular dichroism (CD) and two-dimensional (2D) NMR spectroscopy. Peptide BP1 mediated transamination of pyruvic acid and showed some rate enhancement and optical induction [28]. A family of peptides (CBP01–CBP018, Fig. 10), based on peptide BP1, was designed to investigate the effect of changing the microenvironment surrounding the pyridoxamine functionality. In peptides CBP01 through CBP18 a range of residues were explored at positions 5 and 7, flanking Pam at position 6 (Fig. 10). At position 5, a range of basic residues were included for their potential to interact directly with the pyridoxamine functionality in both the deprotonation and reprotonation steps of transamination. Residue 7, which was not expected to directly interact with Pam, was replaced with amino acids with different electrostatic, steric, and conformational properties to change the relative orientation and environment of Pam and the basic residue at position 5.

These peptides were prepared and purified using a newly developed multiple peptide synthesis and purification strategy [31]. Using this method (Fig. 11), full length peptides could be selectively purified with a biotin capping group. Removal of the biotin group under slightly basic conditions then liberated a peptide capped at the N-terminus with a neutral glycolate group, structurally similar to the commonly utilized acetate capping group.

The ability of peptides CBP01–CBP18 to modulate pyridoxamine-mediated transamination was determined by the conversion of pyruvic acid to alanine in both the absence and presence of copper(II) ion, which would be coordinated by the transamination intermediates [32]. In the absence of copper(II) ion, peptide CBP13 showed up to a 5.6-fold increase in alanine production relative to a pyridoxamine model compound and peptide CBP14 produced alanine with a 27% ee of the l-enantiomer. In the presence of copper(II) ion, peptide CBP13 again showed the greatest increase in product production, with a 31.7-fold increase in alanine production relative to the pyridoxamine model compound. Peptide CBP10 showed optical induction for D-alanine with a 37% ee.

Peptides CBP01–CBP18 also showed interesting trends in optical induction as a function of both reaction conditions and the identity of residues at positions 5 and 7. In the absence of copper(II) ion, L-alanine was the favored stereoisomer product for peptide CBP01–CBP18, while formation of D-alanine was favored in the presence of copper(II) ion. This suggests that the $\beta\beta\alpha$-structure of the pep-

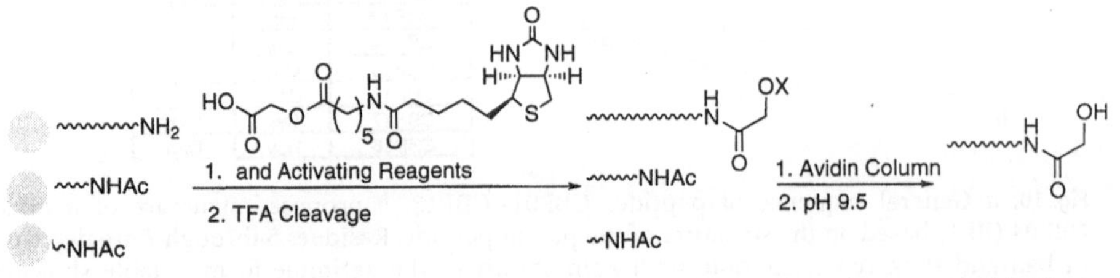

Fig. 11. Purification scheme for full-length glycolate-capped peptides using a reversible biotin capping agent

tide favored different orientations for the metalated and unmetalated reaction intermediates. Furthermore, the identity of the residues at positions 5 and 7 independently influenced the degree of optical induction observed. In the absence of metal ion, the α,γ-diaminobutanoic acid (Dab) residue at position 7 produced a greater degree of optical induction than analogous peptides with either aspartic acid or glycine at this position. In the presence of copper(II) ion, glycine at position 7 generated the greatest optical induction. In the absence of metal ions, bases at position 5 with longer side chains produced greater optical induction. However, in the presence of copper(II) ion, basic residues with shorter side chains produced more optical induction. The observed trends, in addition to the increased product formation, suggest that the $\beta\beta\alpha$-motif is capable of generating a protein microenvironment for the coenzyme group to produce functionally interesting peptides.

3
Thiamine Diphosphate

3.1
Chemical Properties

Thiamine diphosphate (TDP) is an essential coenzyme in carbohydrate metabolism. TDP-dependent enzymes catalyze carbon–carbon bond-breaking and -forming reactions such as α-keto acid decarboxylations (oxidative and non-oxidative) and condensations, as well as ketol transfers (trans- and phospho-ketolation). Some of these processes are illustrated in Fig. 12.

The finding that thiamine, and even simple thiazolium ring derivatives, can perform many reactions in the absence of the host apoenzyme has allowed detailed analyses of its chemistry [33, 34]. In 1958 Breslow first proposed a mechanism for thiamine catalysis; to this day, this mechanism remains as the generally accepted model [35]. NMR deuterium exchange experiments were enlisted to show that the thiazolium C2-proton of thiamine was exchangeable, suggesting that a carbanion zwitterion could be formed at that center. This nucleophilic carbanion was proposed to interact with sites in the substrates. The thiazolium thus acts as an electron sink to stabilize a carbonyl carbanion generated by deprotonation of an aldehydic carbon or decarboxylation of an α-keto acid. The nucleophilic carbonyl equivalent could then react with other electro-

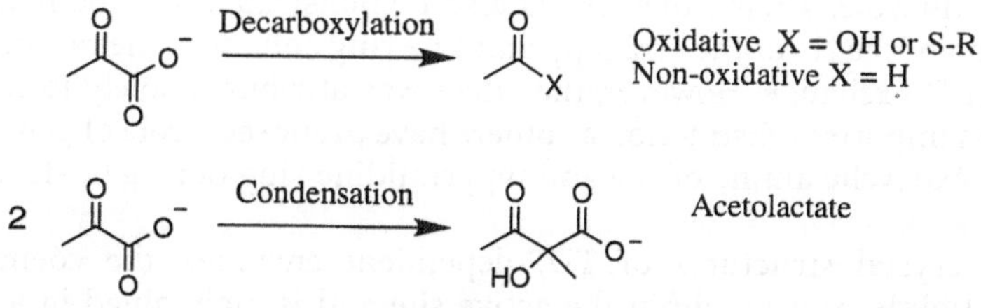

Fig. 12. Reactions catalyzed by TDP-dependent enzymes

philes (for example a proton or α-keto acid) and subsequently detach from the thiazole ring, regenerating the ylide. Thus, thiamine is one of the few coenzymes which requires no recycling or regeneration to complete its catalytic cycle.

The reaction path of thiamine-dependent catalysis is essentially unchanged in the presence of an apoenzyme, except that the enzyme active site residues increase reaction rates and yields and influence the substrate and product specificity. The X-ray crystal structures of TDP-dependent enzymes have clarified this view and permitted an understanding of the roles of the individual amino acids of the active site in activating and controlling the thiazolium reactivity [36–40].

Prior to elucidation of the structures of TDP-dependent enzymes, chemical models and solution studies of the coenzyme predicted roles for the interactions of the apoenzyme with TDP. The influence of a hydrophobic environment on thiazolium reactivity had been clearly shown by both C2-H/D exchange and pyruvate decarboxylation studies. Crosby and Lienhard showed that in ethanol, the rate of C2-H exchange increased by as much as 500-fold and the rate of decarboxylation of pyruvate increased by 10^4- to 10^5-fold [41]. Benzoin and other acetoin condensations have also been found to occur rapidly in organic solvents. It has been proposed that the hydrophobic milieu of the active site is the primary characteristic of the apoenzyme in the acceleration of thiamine catalysis [41]. The crystal structures of transketolase [36] and pyruvate decarboxylase [37] provide significant evidence for these proposals. An obvious feature of the thiamine binding site is the provision of a hydrophobic environment to the coenzyme. In yeast pyruvate decarboxylase, thiamine is almost completely buried in the enzyme active site, with only the S1, C2 and 4'NH$_2$ groups being solvent accessible [37]. The thiazolium and aminopyrimidine rings make extensive van der Waals contacts with hydrophobic active site residues and research by Lobell and Crout suggests that this active site is virtually "waterproof" during the key steps of catalysis [42].

Interactions with the aminopyrimidine ring and the identity of the N3-substituent have also been assigned roles in thiazolium catalysis in solution. Early studies examined the role of the N3-substituent on the catalysis by thiazolium and found that other functional groups at this position supported catalysis, albeit at reduced efficiencies [34]. Trends for the influence of the N3-substituent on C2-H acidity were made clear by Washabaugh and Jencks [43]. Their results showed that H/D exchange rates for thiazolium compounds increased as a result of increased electron-withdrawing inductive effects. Thus a 4-nitrobenzyl-thiazole exchanged faster than a 4-benzylthiazole which exchanged faster than a 4-methylthiazole. Their experiments also demonstrated that methylation or protonation of the N1 of the aminopyrimidine ring enhanced the rate of thiazolium C2-H/D exchange. However, the effect was attributed solely to inductive effects – an important distinction as others have proposed a role of general base for the 4'-exocyclic amine of the aminopyrimidine ring during C2-H deprotonation.

In the crystal structures of TDP-dependent enzymes, the coenzyme is generally tightly packed within the active site and is maintained in a specific conformation which is conserved in all TDP-dependent enzymes but which is energetically unfavorable in the free coenzyme [37]. The binding of the two rings of the coenzyme in a "V" conformation (defined by specific torsion angles between the 2 rings) has therefore become a source of much discussion regarding the relationship between the two rings during the catalytic cycle [44, 45]. Specifically, the role of the aminopyrimidine ring in augmenting or assisting the catalysis has been central to much research. The aminopyrimidine ring makes further contacts with the active site residues through hydrogen bonds – most importantly from a glutamic acid side chain to N1. This hydrogen bond is conserved in all TDP-dependent enzymes and mutations of the glutamic acid to glutamine or alanine severely diminishes the enzyme activity [45].

Enzyme active sites offer several other interactions which help to modulate coenzyme reactivity. The reactivity of free thiamine is sensitive to pH. As with most enzymes, the TDP apoenzymes provide the coenzyme with an environment with differentiated "effective" pH environments. The active sites of these enzymes provide a hydrophobic yet acidic environment to the N1 atom of the aminopyrimidine ring, while maintaining an "effectively" basic environment around the C2-position of the thiazolium ring. Although the base that assists C2-H deprotonation remains a matter of speculation, it has been proposed that it is either a hydroxyl from a bound water or the 4′-amine of the aminopyrimidine ring [42, 45]. These differential "effective" pHs allow the aminopyrimidine ring to remain activated through the Glu–N1 hydrogen bond while the thiazolium and reaction intermediates experience the high pH required to accelerate the reactions.

The complexity of the environment surrounding the coenzyme has prevented most simple model systems from dramatically enhancing thiamine reactivity or specificity [46–48]. Peptide- or protein-based models have the advantage of presenting a reasonably complex environment to the coenzyme functionality within a water soluble, yet synthetically accessible, scaffold.

3.2
Modeling Thiamine Catalysis in Protein and Peptide Systems

As with the pyridoxal coenzymes, modified native protein scaffolds have been used to investigate features of TDP-dependent catalysis. Suckling has generated "thiazolopapain" conjugates by alkylating the active site thiol of papain with 4-bromomethylthiazolium derivatives (3-methyl and 3-benzyl) [49]. The preparation of these derivatives is shown in Fig. 13. The methyl and benzyl thiazolopapains were assayed for pyruvate decarboxylation and found to be catalytically active although no rates or extents of reaction were reported. Suckling also initially reported that the modified enzymes did not perform acetoin or benzoin condensations.

In a subsequent report, however, the thiazolopapains were shown to be competent in catalyzing a carbon–carbon bond-forming reaction of the acetoin condensation type (Fig. 14) [50]. The reaction of the papain derivatives with 6-oxo-heptanal was assayed at neutral pH (Fig. 14). The course of the reaction was monitored by HPLC and the products analyzed by ^{1}H NMR. In the case of the

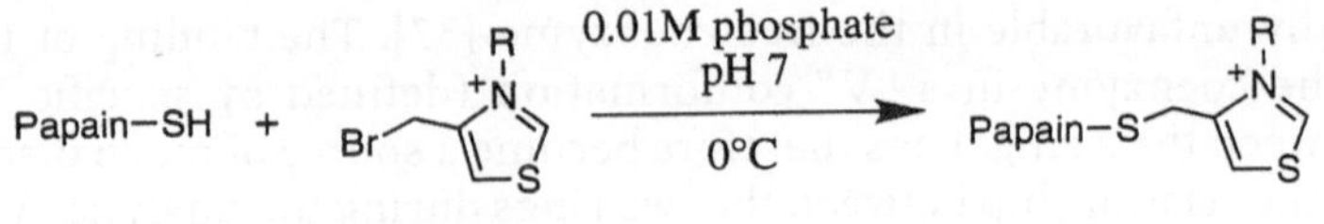

Fig. 13. Preparation of thiazolopapain

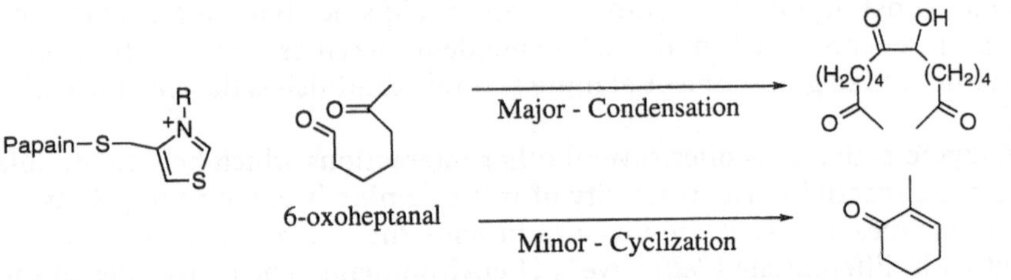

Fig. 14. Thiazolopapain-catalyzed reactions of 6-oxoheptanal

3-benzylthiazolopapain, a near 4000-fold excess of substrate was almost entirely consumed (88%) within 150 h. 3-Methylthiazolopapain was much less efficient, converting only 28% of the substrate in the same time. The products generated included the expected cyclization product as a minor product (8 and 28% for 3-methyl- and 3-benzylthiazolopapain, respectively), and an unexpected condensation product as the major product (20 and 60%, respectively). In contrast, the simple thiazolium model compounds catalyzed the turnover of only 23% of the substrate in a comparable time period. Due to somewhat different reaction conditions, it is difficult to assess the extent of the influence of the papain active site but given the same amount of substrate, a 40-fold smaller amount of thiazolopapain was able to catalyze three times as many turnovers.

Variation in the reactivities of the methyl- and benzylthiazolopapains may be attributed to differences in the inherent reactivities of the methyl- and benzyl-thiazoliums. The rate of consumption of substrate by the benzyl derivative was approximately three times faster than the methyl derivative. Interestingly, model compound studies have shown that the C2-H/D exchange rate of 3-benzyl-4-methylthiazole is three times faster than that of 3,4-dimethylthiazole [43]. It should be noted, however, that Suckling states that it is not possible to say whether the reactions occur within the active site pocket of papain or on the surface of the protein. Therefore it is not clear whether a hydrophobic environment is being fully exploited in this system. This system has not been further developed nor has any structural data been garnered to attempt to explain the role of the protein in modulating the thiazolium activity. However, discrete interactions with the papain architecture could be envisioned and information from X-ray structures used as a basis for computer modeling to predict or design interactions between the coenzyme functionality and the "active site".

3.3
Thiazolium Amino Acid Chimera Peptides

Thiazolium chemistry has also been explored within the context of designed peptide scaffolds. In this case the extent to which a relatively small host architecture can influence thiazolium reactivity has been challenged. Recently, a number of research groups have developed small (< 50 amino acids) polypeptide motifs with defined secondary and supersecondary structures [51]. These peptides could potentially be modified with a thiazolium functionality by post-synthetic chemical modification; however, maintaining the coenzyme in a rigid orientation presents a major challenge in such small systems. Consequently, alternate strategies have been employed. In research patterned after the pyridoxal and pyridoxamine amino acid chimeras, the Imperiali group has adopted a parallel strategy for the investigation of thiamine-dependent processes.

A thiazolium amino acid (Taz) has been developed which can be utilized to mimic TDP-dependent enzyme function [52]. In this strategy, illustrated in Fig. 15, the commercially available amino acid 4-thiazolylalanine is incorporated into peptides by solid phase peptide synthesis. Prior to deprotection of the amino acid side chains and cleavage of the peptide from the resin, the thiazole amino acid is alkylated with an alkyl halide to generate the corresponding thiazolium amino acid having various N3-substituents (BzTaz = 3-benzyl-Taz, NBTaz = 3-nitrobenzyl-Taz).

Taz has been incorporated into peptides which have been designed to adopt varying degrees of peptide secondary and supersecondary structure (shown in Table 3) [53]. Peptide BzTaz-hairpin is an octapeptide which is designed to include a constrained type II' β-turn. This peptide is designed to present the BzTaz on one face of an antiparallel β-sheet, surrounded by tyrosine 1, valine 3 and phenylalanine 8. The Telix1 peptide is based on the DeGrado α1b peptide [54] which is designed to assemble into a four-helix bundle. The placement of NBTaz at position 6 was intended to situate the thiazolium ring at the interface of the hydrophobic core of leucine residues rather than to integrate it directly into the core where it could destabilize the helix association. The secondary structures of these peptides were confirmed by CD spectroscopy. Spectroscopic studies are consistent with the assigned β-turn structure for BzTaz-hairpin, while the Telix peptide exhibits a CD signature reminiscent of helix bundle peptides, including concentration-dependent helicity.

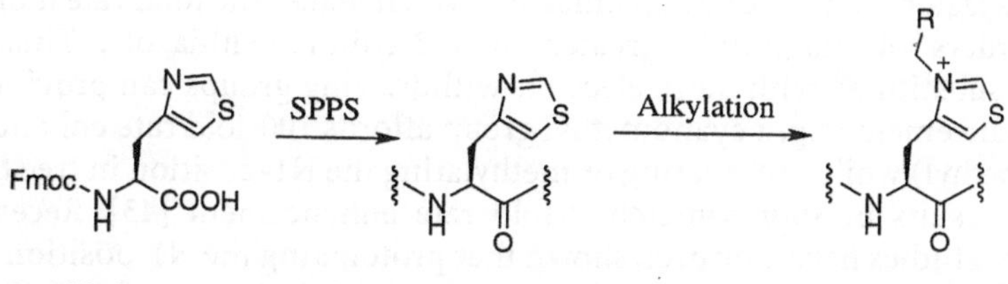

Fig. 15. Peptide incorporation of thiazolium amino acid chimera

Table 3. Taz peptide sequences

Name	Sequence	Predicted structure
BzTaz-hairpin	Ac-Tyr-Arg-Val-DPro-Ser-**Taz(Bz)**-Asp-Phe-NH$_2$	Type II' hairpin
NBTelix1	Ac-Ser-Ala-Leu-Glu-Glu-**Taz(NB)**-Leu-Lys-Lys-Leu-Ala-Glu-Leu-Leu-Lys-Gly-NH$_2$	Four-helix bundle (tetrameric)

Table 4. Relative H/D exchange kinetics of C2-proton of Taz peptides and compounds in 0.2 M d_4-acetate buffer in D$_2$O, pD 4.7

Entry	Substrate	Relative rate	Structure effect
1	3,4-dimethylthiazole	1.00	-
2	3-benzyl-4-methylthiazole	3.41	-
3	3-nitrobenzyl-4-methylthiazole	8.90	-
4	BzTaz-hairpin	17.9	5.25
5	peptide NBTelix	1110.3	125

The effect of the peptide environment on the reactivity of the Taz amino acid was examined by determining the rate of H/D exchange of the thiazolium C2-proton by ^{1}H NMR. The exchange rates of these peptides, relative to 3,4-dimethylthiazole, are presented in Table 4. Exchange rates of each peptide relative to the respective model compounds (e.g. BzTaz-hairpin vs. 3-benzyl-4-methylthiazole) are intended to highlight the contribution of the peptide architecture to the observed exchange rate and are presented under "Structure Effect" in Table 4. The H/D exchange data show striking signs of the influence of a complex peptide architecture on the activation of the thiazolium ring. The β-hairpin peptide exhibits a small influence on the acidity of the thiazolium C2-proton while the Telix scaffold enhances the rate of H/D exchange by over two orders of magnitude.

The influence of the N3-substituent has also been shown to be useful in modulating Taz H/D exchange. As detailed earlier, the more electron-withdrawing substituents enhance the rate of C2-deuterium exchange [43]. In the Taz system, a nitrobenzyl substituent affords NBTelix an approximate 10-fold rate enhancement in H/D exchange rate (3-nitrobenzyl-4-methylthiazole exchanges 8.9 times faster than 3,4-dimethylthiazole) which makes the total rate more than three orders of magnitude greater than 3,4-dimethylthiazole. Thiazolium systems substituted with more electron-withdrawing groups can provide more rate enhancement (e.g. a cyanomethyl group affords 100-fold rate enhancement over a methyl) while protonating or methylating the N1-position in free thiamine only results in approximately 3-fold rate enhancement [43]. Recent H/D exchange studies have, however, shown that protonating the N1-position within the active site of yeast pyruvate decarboxylase can be responsible for significantly enhancing H/D exchange rates (~ 350-fold) [45]. Whether the Taz systems will be as effective at modulating thiazolium-catalyzed reactions has yet to be investigated. However, that simple thiazolium salts placed in hydrophobic solvents show enhanced reactivity is indicative of the promise in this approach.

More complex interactions present immense challenges for model systems. Creating complex microenvironments within an active site for a thiazolium catalyst where specific hydrogen bonds and electrostatic interactions are formed would not be possible without the modularity and diversity of the side chains of a peptide-based system. Peptide-based modeling of TDP-dependent enzymes, as with the Taz amino acid chimera approach, may prove amenable to creating distinct interactions within a designed active site. The coenzyme chimeras and other peptide-based approaches depend on well-designed, structurally characterized and stable peptide scaffolds. Detailed structural analysis of first-generation designs can now guide in the design of "next generation" systems which may allow for the inclusion of specific interactions to further alter the thiazolium.

4
Flavin Coenzymes

4.1
Chemical Properties

The flavin-based coenzymes flavin adenine dinucleotide (FAD) and flavin mononucleotide (FMN) are associated with a wide variety of enzymes that catalyze reactions in critical biosynthetic and catabolic processes (Fig. 16). Unlike other coenzymes, the reactions catalyzed do not conserve specific mechanistic pathways. In each case the apoenzyme serves to steer the course of the reaction through specific interactions with substrate and coenzyme [55]. Nonetheless, there are common features of the interactions of the apoenzymes with the flavin which can be exploited in the design of functional peptides and proteins.

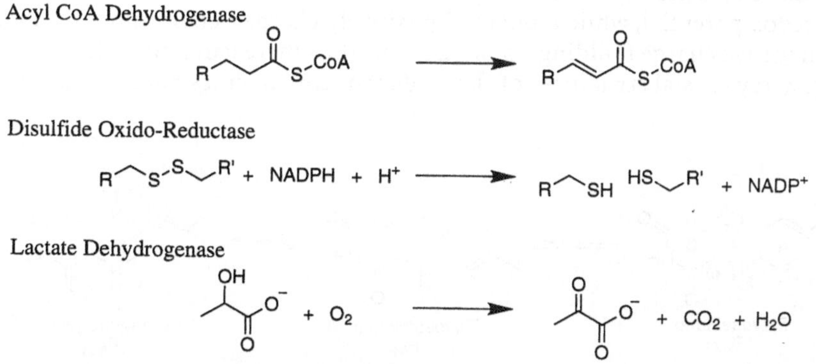

Fig. 16. Examples of flavoprotein-catalyzed reactions

FAD and FMN are capable of mediating both one- and two-electron oxidations and reductions, unlike the nicotinamide-based redox coenzymes. Additionally, the flavins differ from nicotinamide coenzymes in their association with the corresponding apoenzymes. Most flavins are tightly bound within the protein and a few are covalently bound. Thus the recycling of these coenzymes occurs through interactions with a second substrate; in fact, flavin-dependent enzymes are frequently classified by their regenerative half reactions [56]. Reduced flavins are rapidly reoxidized by molecular oxygen, a characteristic which allows them to be easily recycled as catalysts in model systems. The core functionality of the flavin coenzymes rests in the isoalloxazine ring system. The N10-appendages are generally restricted to binding functions, although in some cases they are involved in the reaction mechanism [57]. The isoalloxazine system features three rings which give the flavin an amphipathic character. The hydrophobic dimethylbenzene ring designated A is in contrast to the more polar pyrimidine C ring (Fig. 17). The two are joined by the phenylenediamine ring B through which electrons are delivered to or removed from the flavin.

The flavin functionality has three important redox states relevant to enzymatic catalysis [58]. The most oxidized form of the isoalloxazine ring is the flavoquinone or Fl_{ox}. Reduction of Fl_{ox} by a single electron yields the resonance-stabilized semiquinone radical or Fl_H. The fully reduced isoalloxazine ring is designated the flavohydroquinone form or Fl_{red}. The different forms of the isoalloxazine ring have obvious electronic but also steric differences which alter their reactivity or which can be affected by interactions with an apoenzyme. For example, the Fl_{ox} ring system is relatively rigid and planar while the Fl_{red} form is kinked (the B ring is no longer planar). Binding to an apoenzyme could sterically influence the equilibrium between these two forms by enforcing a more or less planar conformation on the flavin [56]. Interactions with different rings of the flavin has also been shown to modulate activity. Chemical models have shown that tethering electron-withdrawing groups such as a cyano to the A ring enhances the electron deficiency of the isoalloxazine ring and increases the reduction potential [59]. The polarity of the environment around ring C can also greatly affect the redox potential of the coenzyme. Flavins tethered to cationic polymers have been shown to oxidize substrates which show no reactivity with free flavins [59]. A negatively charged or hydrophobic environment will decrease the redox potential, while a polar or positively charged environment stabilizes the negative charge building up on the ring, thus increasing the redox potential [55]. X-ray crystal structures of flavin-dependent enzymes have confirmed the

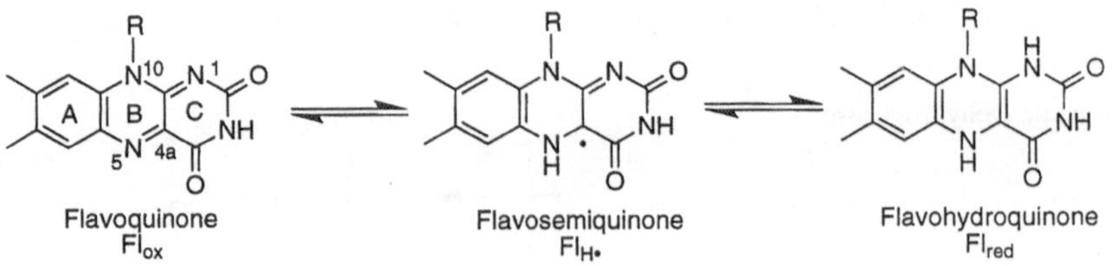

Fig. 17. Flavin redox states

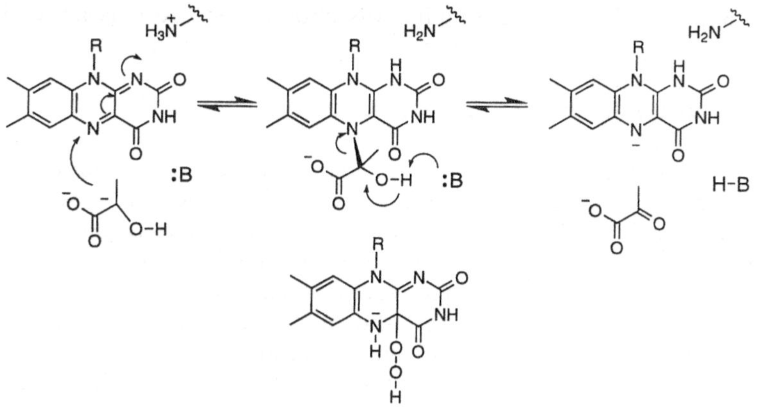

flavin 4A-hydroperoxide

Fig. 18. Covalent substrate-coenzyme adducts in flavin biochemistry

presence of groups which interact with ring C in this fashion to modulate the flavin redox potential. The magnitude of these effects is illustrated by the range of the reduction potentials observed in flavoproteins D from – 495 to + 80 mV (a ~ 600 mV or 14 kcal/mol range) [55].

The kinetics of the reactions of substrates with flavins can also be modulated by the apoenzyme. As previously indicated, the mechanistic pathways of flavo-protein-catalyzed reactions are not unique. Some enzymes catalyze a reaction sequence where the substrate becomes covalently tethered to the flavin, rather than simply receiving a transferred hydride. For example as shown in Figure 18, it has been proposed that l-lactate dehydrogenase forms an adduct between the flavin N5 and a carbanion generated alpha to the lactate carboxylate. The α-hydroxyl can be deprotonated and the flavin released and reduced to yield the pyruvate product [55]. A second example is the reoxidation of Fl_{red} by oxygen. In this case, the oxygen substrate forms an adduct with the flavin at the C4a-position. The apoenzyme can therefore utilize steric effects to guide a reaction towards or away from a specific pathway [55]. This level of control over the reactivity and reaction selectivity of the flavin functionality is a major challenge for the design of functional mimics of flavoproteins.

4.2
Modeling Flavin Coenzyme Function in Peptides and Proteins

Kaiser pioneered the introduction of coenzyme functionality into protein scaffolds in 1977 [60]. He introduced a strategy of chemical modification to incorporate the functionality into or adjacent to an existing enzyme active site. The goal of this strategy was to take advantage of the substrate specificity of the active site to effect the chemistry of the coenzyme functionality. This research incorporated a flavin group into four different enzyme systems: papain, glyceraldehyde-3-phosphate dehydrogenase (GADPH), lysozyme, and hemoglobin [61–64]. The flavin functionality was chosen because it is active in the absence of an apoenzyme. The rationale was, therefore, to utilize the binding site of the enzyme to orient the substrate in a favorable fashion, as opposed to attempting to create specific interactions between the flavin and functional groups of the enzyme.

The first and most extensively examined system was the hydrolytic enzyme papain. A variety of isomeric α-bromoacetylisoalloxazines were used to selectively tether a flavin moiety to the active site cysteine residue. Different isomeric linkages were proposed to allow orientations of the flavin relative to the substrate binding site which would favor reactions with a bound substrate [65].

These "flavopapains" (Fig. 19) were shown to be effective redox catalysts for the oxidation of N-alkyl-1,4-dihydronicotinamides. The localization of the flavin moiety adjacent to the hydrophobic binding groove of the active site further allowed the constructs to exhibit substrate selectivity and, in some cases, saturation kinetics. The most effective flavopapain was the 8-isomer (FP-8) which reacted rapidly with a variety of N-alkyl-1,4-dihydronicotinamides. The best substrate was N-hexyl-1,4-dihydronicotinamide for which the k_{cat}/K_m of its oxidation by FP-8 was determined to be in the vicinity of 10^6 M^{-1}s^{-1} in air saturated buffer at pH 7.5 and 25 °C. This represented an approximately 10^3-fold rate enhancement over the simple flavin model compound.

FP-8 also exhibited substrate selectivity, oxidizing N-benzyl-1,4-dihydronicotinamide approximately 200-fold faster than the model. NADH was a very poor substrate for FP-8, which effected a 4-fold rate enhancement in its oxidation. This is consistent with the hydrophobic binding site of papain playing a role in binding these substrates adjacent to the flavin. NADH, having a relatively hydrophilic N1-substituent, binds less effectively than the N-benzyl- or hexylnicotinamides. The ability of flavopapain FP-8 to achieve these impressive rate enhancements is in contrast to the next best flavopapin, FP-7, which achieved only 30-fold rate enhancement over the corresponding model compound for the reduction of N-benzyldihydronicotinamide. Even more striking was the lack of any significant rate enhancement with FP-6.

The α-acetyl group of the flavin–papain linkage was proposed to play a role in making the 8-acetylisoalloxazine papain a more effective catalyst. Computer modeling suggested that the carbonyl oxygen, if aligned properly, could hydrogen bond to both the backbone amide of cysteine 25 (to which the flavin is te-

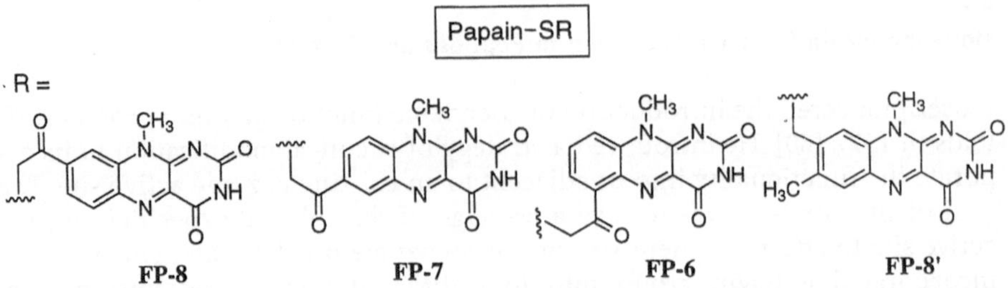

Fig. 19. Designed flavopapains

thered) and to the side chain amide of glutamine 19. The differing activities of FP-7 and FP-8 are attributed to this hydrogen bond aligning the 8-isomer in a more favorable orientation that allows efficient interaction with a substrate bound in the hydrophobic binding pocket of papain. Analysis of FP-8′, which lacks the acetyl carbonyl oxygen, supports this hypothesis as it effects only a 3-fold rate enhancement over its model compound and did not exhibit saturation kinetics. Also, computer modeling of FP-6, which exhibited no rate enhancement, suggested that hydrogen bonding of the acetyl group to the papain amides would require the isoalloxazine ring to swing away from the substrate binding site, preventing the flavin from taking advantage of the papain active site.

The Kaiser approach highlighted the power of enhancing coenzyme reactivity through incorporation into alternative protein scaffolds. These highly efficient semisynthetic systems derived catalytic potential primarily through the organization of the flavin moiety relative to the substrate binding site as opposed to the arrangement of protein-bound functional groups which could perform specific portions of the reaction, stabilize a high energy intermediate, or enhance the inherent reactivity of the coenzyme. However, the interaction of the papain with the 8-acetylisoalloxazine carbonyl demonstrated that specific interactions between the coenzyme and the protein could be achieved and, with greater control over the primary sequence of the protein scaffold through site directed mutagenesis or chemical synthesis, more precise and influential interactions could be envisaged.

4.3
Investigation of Flavin-Modified Peptides

In recent research, Nishino and co-workers have incorporated a flavin functionality into synthetic, *de novo* designed four-helix bundles. The first system to be investigated was a synthetic 53-residue four-helix bundle peptide prepared via solution condensation of three 13-residue peptides and one 14-residue peptide [66]. A flavin moiety was introduced by alkylation of a cysteine residue at position 8 of the first helix with 7-bromoacetyl-10-methylisoalloxazine. Nishino has also reported another four-helix bundle containing a flavin moiety. In this case, the four helices were tethered to a cyclic pseudo-octapeptide with alternating helices attached *N*- or C-terminally [67]. The resulting four-helix bundle would be assured an antiparallel configuration with the flavin functionality attached to position 7 of the two C-terminally attached helices. Both peptides maintained a high degree of helicity, theoretically including the flavin(s) within the hydrophobic core of the four-helix bundle.

The ability of this hydrophobic peptide environment to effect the flavin reactivity was assessed by monitoring the oxidation of *N*-alkyl-1,4-dihydronicotinamides (*N*-alkyl-NAH) using conditions similar to those reported in the Kaiser experiments. Under these conditions, both systems exhibited slight enhancements in the rate of oxidation of either *N*-benzyl-NAH or *N*-hexyl-NAH. Hypothesizing that the lack of performance of the system may have been due to the occlusion of the flavin from interacting with the substrate, the assays were repeated in the presence of sodium dodecyl sulfate (SDS) to "loosen" up the helix bundles. Under these slightly denaturing conditions, a 4-fold rate enhancement was observed with the 53-residue bundle for both substrates, while the templated bundle "pseudoprotein" afforded approximately a 16-fold rate enhancement for the oxidation of N-hexyl-NAH (8-fold per flavin).

The Nishino studies represented the first attempt to incorporate flavin functionality into a synthetic and structured polypeptide motif. As with the Kaiser systems, it is possible that the observed rate enhancements are a result of the association of the hydrophobic substrates with the hydrophobic core of the bundles. The use of SDS to increase accessibility to the flavin would seem to support this. However, it is also possible that the placement of the flavin into the hydrophobic core has decreased its reduction potential and that the SDS interacts with the bundle to create a slightly more hydrophilic environment for the flavin while preserving some of the hydrophobicity for substrate binding.

4.4
Flavin Amino Acid Chimeras

Carell has recently presented the study of a flavin amino acid chimera to model riboflavin in DNA photolyases [68]. This amino acid L1 (Fig. 20) was synthesized in an enantiopure fashion by building the alloxazine ring onto the epsilon amine of lysine. This coenzyme chimera was applied to the problem of repairing DNA damage caused by UV irradiation. L1 was incorporated into an 21-residue peptide, P-1, possessing the sequence of the DNA-binding domain of the helix-loop-helix transcription factor MyoD.

The ability of peptide P-1 to repair cyclobutane uracil dimers was investigated. The peptide was incubated with a synthetic oligonucleotide containing a cyclobutane uracil dimer with formacetal linkages rather than natural phosphodiester bonds. This oligonucleotide had been shown to be a substrate for natural photolyases. Photoreduction of the isoalloxazine of L1 in P-1 was accomplished in the presence of substrate oligonucleotide under anaerobic conditions with EDTA and 366 nm light or daylight. This resulted in the complete conversion of the lesion-containing oligonucleotide to the repaired oligonucleotide, as monitored by HPLC. The irradiation time required for complete conversion varied

R-Ala-Asp-Arg-Arg-Lys-Ala-Ala-Thr-Glu-Arg-Glu-Arg-Arg-Arg-Xaa-Ser-Lys-Val-Asn-Glu-Ala-NH$_2$

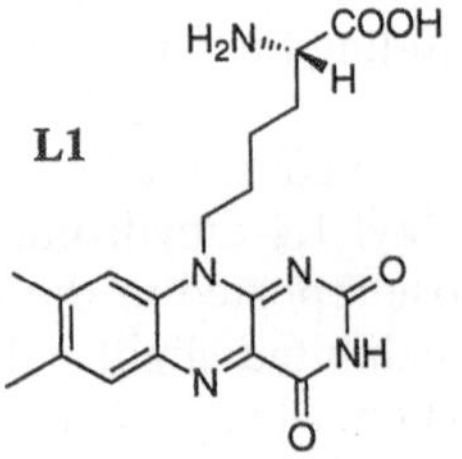

P-1; R = Ac, Xaa = L1

P-2; R = Ac, Xaa = Leu

P-3; R = Cys-Gly-Gly, Xaa = L1

P-4;

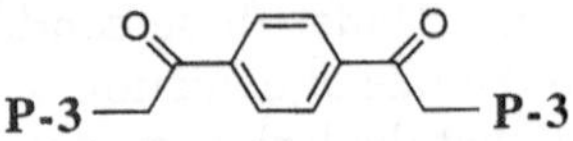

Fig. 20. Structures of flavin amino acid chimera L1 and peptides P-1 – P-4

depending on the concentration of peptide and oligonucleotide, as well as the ionic strength of the medium. Carell explains that this indicates an ionic association of the positively charged residues of the peptide with the phosphodiester backbone. The peptide–oligonucleotide complex is most likely unstructured since the transcription-factor elements require double-stranded DNA for structurally well-defined complexes. Control studies with peptide P-2, in which a leucine replaces L1, indicated no dimer repair. Preliminary results with P-4, a dimer of P-3, indicate that this peptide construct was able to catalyze the repair of the oligonucleotide faster than P-1, but at a lower stoichiometry (in the vicinity of 20 mol%). These results indicate the importance of the peptide architecture to allow localization of the flavin moiety to the oligonucleotide. This association is likely non-specific; however, the flavin amino acid chimera could be incorporated into small peptides with a more precise, even sequence specific, DNA binding to more efficiently localize the coenzyme functionality to the oligonucleotide.

5
Nicotinamide Coenzymes

5.1
Chemical Properties

The nicotinamide coenzymes nicotinamide adenine dinucleotide (NADH) and nicotinamide adenine dinucleotide phosphate (NADPH) are associated with a wide variety of enzymes involved in oxidation-reduction reactions (Fig. 21). NADH is typically involved in oxidative catabolic reactions, while NADPH is primarily used in biosynthetic pathways [58].

Distinct coenzymes are required in biological systems because both catabolic and anabolic pathways may exist within a single compartment of a cell. The nicotinamide coenzymes catalyze direct hydride transfer (*from* NAD(P)H or *to* NAD(P)$^+$) to or from a substrate or other cofactors active in oxidation-reduction pathways, thus acting as two-electron carriers. Chemical models have provided

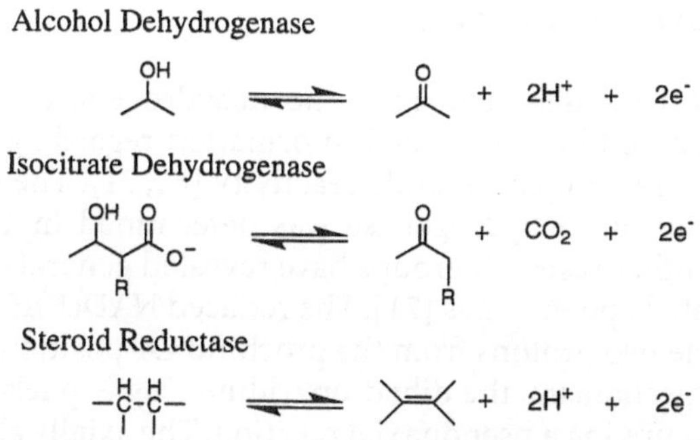

Alcohol Dehydrogenase

Isocitrate Dehydrogenase

Steroid Reductase

Fig. 21. Examples of reactions catalyzed by NAD(P)H-dependent enzymes

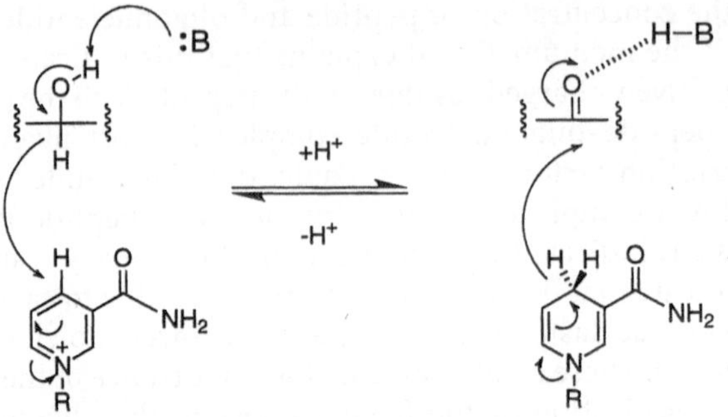

Fig. 22. Mechanism of nicotinamide-dependent redox reactions

important insights into the reactivity of nicotinamide-based hydride transfer systems [59]. Some of these have achieved reactivities and stereospecificities comparable to enzyme systems, thus revealing possible influences of the apo-enzyme on the natural coenzyme.

The mechanism of hydride transfer by the nicotinamide coenzymes generally follows the scheme shown in Fig. 22 [58]. This mechanism highlights the importance of pH on the redox reaction, regardless of whether the reaction occurs free in solution, in an enzyme active site, or in a designed peptidyl model system. A proton is released during oxidation and a proton is correspondingly consumed during reduction. Thus, oxidation is favored at high pH, while more acidic conditions enhance the reduction reaction [69]. However, at either extreme, the nicotinamide pyridinium and dihydropyridine species are each susceptible to either base- or acid-catalyzed decomposition [70].

The ionization state of the coenzyme is also important. During reduction a charged pyridinium species is created while during oxidation the charge is lost. Thus, more polar environments favor reduction while more hydrophobic conditions favor oxidation [69]. Therefore the apoenzyme environment and model system scaffolds must not only enhance the reactivity of the coenzyme, but must also address these issues of equilibrium and stability.

5.2
Nicotinamide Coenzymes in Enzymes

Of the NAD(P)H-dependent enzymes, the dehydrogenases have been most extensively studied and have provided information regarding the influence of apoenzyme structure on nicotinamide reactivity [69, 71]. The crystal structure of the unligated lactate dehydrogenase was determined in 1973. Since then, efforts by several other research groups have revealed general characteristics of NAD(P)H associated apoenzymes [71]. The reduced NAD(P)H dihydropyridine transfers one of the two protons from the prochiral C4-position to the substrate. In the proposed mechanism, the dihydropyridine ring is puckered, with one of the C4-protons occupying a pseudoaxial position. The axially aligned C–H bond is proposed to be weaker and thus it is this hydride that is transferred specifically to the substrate. NMR was first used to identify a non-equivalence in the two protons, and this phenomenon has since been supported experimentally and computationally [69]. It has also been shown that dehydrogenases and other NAD(P)H-dependent enzymes are stereospecific in the hydride transfer, transferring to or from only one of the prochiral positions on the dihydropyridine ring [69]. This selectivity comes from the asymmetry of the coenzyme binding site, where one face of the nicotinamide is shielded from the substrate.

Experiments with free NAD$^+$ and lactate dehydrogenase have suggested that the apoenzyme binding site provides a medium of decreased polarity, effectively increasing the electrophilicity of the pyridinium ring which would favor oxidation [69]. Similarly, dihydrofolate reductase, which performs a reduction reaction, features a much more polar NADPH binding site. By binding more tightly to the "product" form of the coenzyme, a given enzyme can use the binding energy to shift the equilibrium of the reaction to favor products, even when the solution reaction equilibrium favors reactants. With this tighter binding, release of the product coenzyme can become rate limiting.

A further role of the "solvation" of the coenzyme by the apoenzyme binding site is to shield the coenzyme against pH-mediated decompositions. Moreover, provision of active site residues to "activate" the substrate permits the oxidation or reduction reaction to proceed at physiological pH where the decomposition reactions are minimized. For example, the reduction of a carbonyl group would utilize an enzyme-associated electrophilic catalyst which may be an acidic amino acid side chain or a metal ion. Other chemical groups covalently attached to the nicotinamide also influence the nicotinamide reactivity through interactions with the apoenzyme. While the adenine dinucleotide is not *directly* involved in the mechanism of hydride transfer, it does provide the binding energy required to enforce the stereoselectivity and to perturb the reaction equilibrium towards products. Finally, the C3 carboxamide is also capable of participating in binding interactions. This group can interact with the apoenzyme and orient the pyridine ring in a favorable arrangement for catalysis [72].

5.3
Modeling Nicotinamide Coenzyme Function in Protein and Peptide Systems

The first report of the incorporation of a nicotinamide functionality into a protein scaffold came from Suckling and co-workers in 1993 [49]. Using methodology developed earlier by Kaiser, nicotinamide-modified papains were generated by treatment of the thiol protease with 3-bromoacetyl-N-benzylpyridine or 3-bromoacetyl-N-ethylpyridine (Fig. 23). Initial studies with these pyridinopapains examined their ability to catalyze the reduction of the activated carbonyl groups of trifluoroacetophenone and ethyl pyruvate. These studies demonstrated that the pyridinopapains were modestly reactive and yielded slight enantiomeric excesses in product distributions. In particular, N-benzylpyridinopapain was able to reduce ethyl pyruvate in 96% yield with sodium dithionite present to regenerate the dihydropyridine form of the pyridinopapain. Under similar conditions, N-ethylpyridinopapain only reduced 6% of ethyl pyruvate. In the presence of N-benzyl-1,4-dihydroniciotinamide as the recycling agent, N-ben-

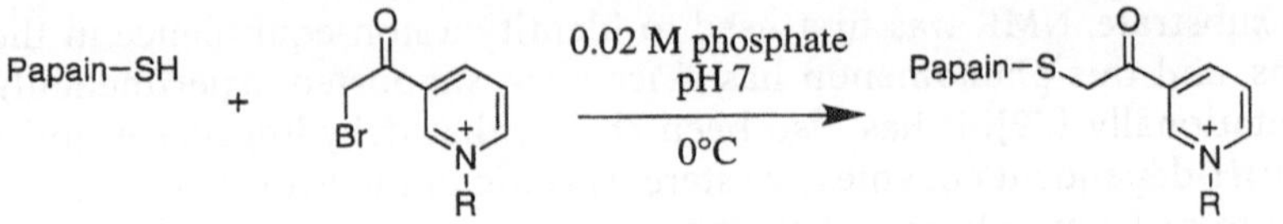

Papain—SH + [structure: Br—CH₂—C(=O)—pyridinium with N⁺—R] → 0.02 M phosphate pH 7, 0°C → Papain—S—CH₂—C(=O)—pyridinium with N⁺—R

R = ethyl or benzyl

Fig. 23. Preparation of pyridinopapains

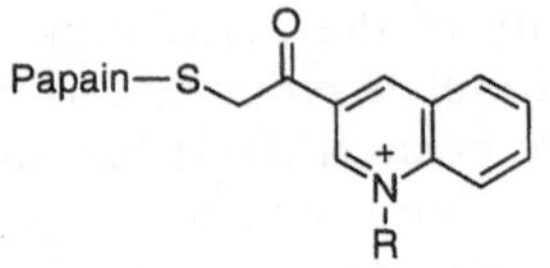

Papain—S—CH₂—C(=O)—quinolinium with N⁺—R

R = methyl, benzyl

Fig. 24. Quinolinopapain

zylpyridinopapain reduced trifluoroacetophenone in only 55% yield, but did so with 11.6% ee (R). The recycling agent alone was able to reduce the substrate in 55% yield, but in this case the reaction was not stereoselective. N-Ethylpyridinopapain was able to reduce trifluoroacetophenone in 71% yield with 14.2% ee (R). The pyridinopapains were not able to reduce unactivated carbonyl compounds such as benzaldehyde or cyclohexanone under these conditions.

Subsequent to these studies, Suckling reported a unique reaction of the pyridinopapains with pyruvic acid [73]. The inability of the pyridinopapains detailed above to provide significant catalysis over and above the recycling agent alone suggested that there may have been competing side reactions. One possibility was that dihydropyridine group of the pyridinopapains acted as an enamine to form an adduct with electrophiles at C5. To exclude this possibility quinolinepapain (Fig. 24) was prepared, a derivative which should presumably preclude such reactions. This modified catalyst reacted in a similar fashion to the pyridinopapains with pyruvate. This led the researchers to propose an alternate, unusual side reaction whereby the carbonyl of the dihydropyridino or quinolinepapain reacted with the pyruvate to generate a covalent adduct.

These authors point out that these results emphasize the difficulty in achieving specific function with this system and highlight the necessity for very precise interactions between the coenzyme functionality and its environment to control and augment reactivity. Peptidyl systems may offer more control over the coenzyme environment as they are readily synthetically accessible and easily manipulated. However, generating stable peptidyl structures presents a significant design challenge.

5.4
Nicotinamide Coenzyme Amino Acid Chimeras

The Imperiali group has employed the coenzyme amino acid chimera approach to construct model systems of NADH-dependent processes [52]. The amino acids utilized for peptide synthesis incorporate a 3-carboxamido pyridine (nicotinamide) group covalently attached via a side chain amide to the side chain of lysine. The pyridine ring is subsequently quaternized through a post-synthetic alkylation to yield the NADH coenzyme chimera Lys(Nic). This methodology was also used to generate the residue Dap(Nic) in which the tetra-methylene lysine side chain (n = 4) was replaced with a shorter β-aminoalanine (n=1) to allow less conformational flexibility of the coenzyme functionality within a peptidyl template [74] (Fig. 25).

The Lys(Nic) and Dap(Nic) residues were incorporated into a simple hexa-peptide structure which contained a constrained type II β-turn (Fig. 26). Dithionite reduction of both Lys(Nic) and Dap(Nic) peptides rapidly generated

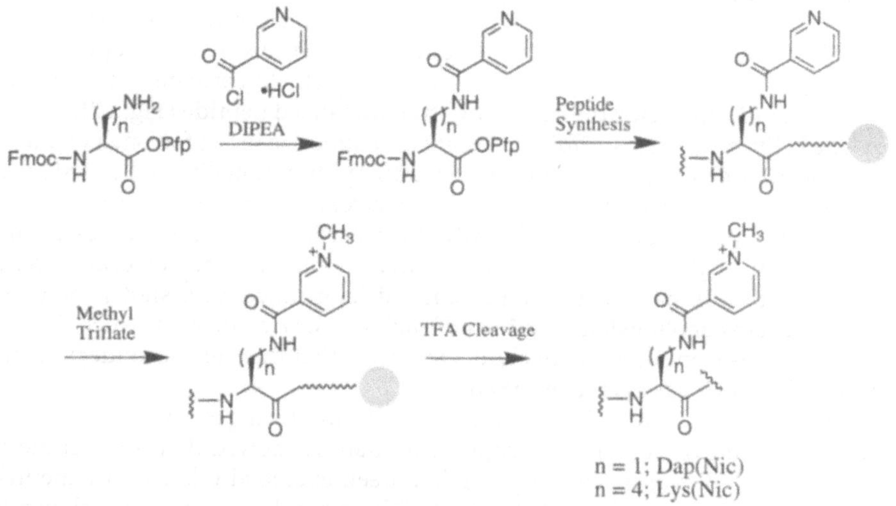

n = 1; Dap(Nic)
n = 4; Lys(Nic)

Fig. 25. Nicotinoyl amino acid chimera synthesis and peptide incorporation

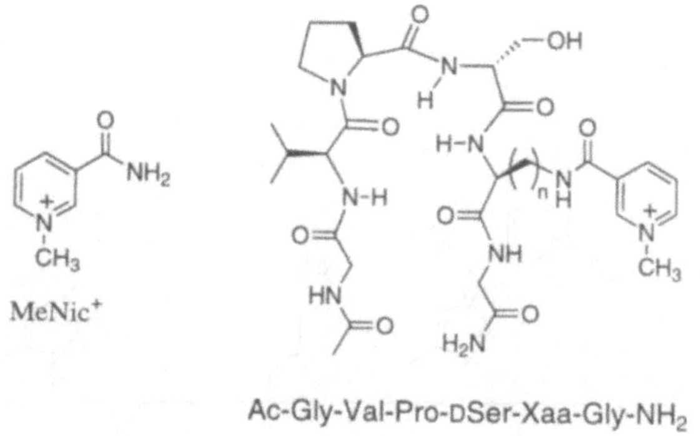

Ac-Gly-Val-Pro-DSer-Xaa-Gly-NH$_2$

Fig. 26. Compounds (in oxidized form) analyzed for the stability of the corresponding dihydropyridines

the dihydropyridine species. There was, however, no significant increase in the lifetime of this species relative to the free methyl nicotinamide (MeNic). CD and NMR analyses of the Lys(Nic) peptide before and after pyridine quaternization showed that β-turn characteristics were lost upon generating the pyridinium derivative [74]. Therefore, the short lifetime of the dihydropyridine species is not surprising given the lack of a stable peptide structure to protect the reactive intermediate. However, these initial studies suggest that a more structured peptide may be capable of more effectively stabilizing the dihydropyridine species.

Most recently, Baltzer and co-workers have incorporated a lysine-bound nicotinamide into a more complex peptide scaffold [75]. This approach takes advantage of the augmented reactivity of a lysine residue contained in a helix-turn-helix scaffold (as described previously [76]). An adjacent histidine is able to selectively catalyze the formation of an amide bond between activated esters and the lysine ε-amino group under aqueous conditions. Thus, reaction of the 42-residue peptide LA-42 with p-nitrophenyl N-methylnicotinate in an aqueous solution at pH 5.9 yields the nicotinoyl-functionalized peptide (Fig. 27).

The helix-turn-helix scaffold is designed to dimerize into a four-helix bundle. Modification of the peptide with the nicotinoyl functionality did not significantly perturb the peptide structure. CD spectroscopy showed no loss in helicity from the parent peptide and NMR spectroscopy confirmed successful incorporation of the nicotinoyl group as well as maintenance of crucial NOE connectivities. In particular, the presence of long-range NOE signals between the side chains of phenylalanine 38 and leucine 12 or isoleucine 9, which lie near the C- and N-termini, respectively, demonstrate that the supersecondary structure of the motif has been conserved.

Reduction of the nicotinoyl group with dithionite at pH 7 was followed by both NMR and UV/vis spectroscopy. This analysis showed that the lifetime of the generated dihydropyridine group had been extended relative to 1-methyl-nicotinamide by more than a factor of 3. This system has not yet been shown to be capable of supporting nicotinamide-catalyzed reduction of carbonyl compounds; however, the successful inclusion of this functionality within a stable

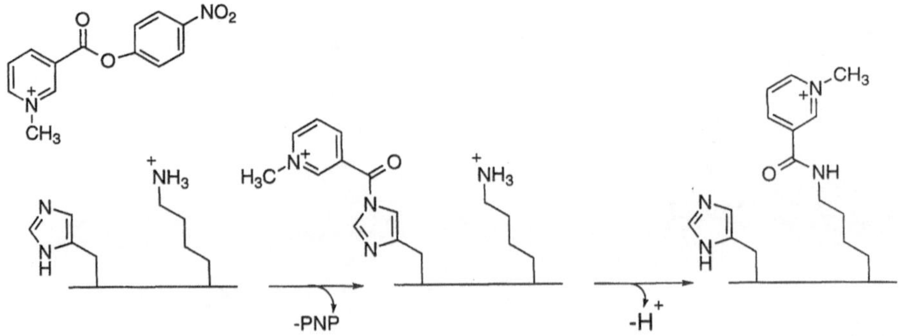

Fig. 27. Baltzer preparation of nicotinoyl-modified LA-42

peptide motif has addressed the primary issues involved in generating a stable dihydropyridine in an aqueous environment. Such systems should allow the design of specific interactions between the peptide scaffold and the nicotinamide as well as elements for substrate recognition and substrate activation.

6
Perspectives

The inherent chemical reactivity of organic coenzymes has attracted a great deal of attention over the past fifty years. The investigation of coenzyme function initially focused on defining the chemical nature of these biotic constituents and cataloging their involvement in enzyme-catalyzed processes. This research was complemented by detailed physical organic and mechanistic studies that interrogated the mode of action of the various coenzymes. Through these studies researchers began to appreciate that while the coenzymes in isolation manifest interesting intrinsic reactivity, it is only when bound to cognate protein hosts that full reactivities and selectivities are realized. Still more remarkable is that the same coenzyme can "channel" a variety of reactions depending on the protein host in which it resides. Subsequently, an area of significant interest involved the development of model systems to emulate the activities of coenzyme-dependent enzymes. These studies were initiated to provide information on the respective roles of the coenzyme and protein functionality in specific transformations. Initial efforts focused on the use of relatively small organic host molecules such as the cyclophanes and cyclodextrins. However, while such minimalist model systems provided interesting insight, it became evident that these constructs might not provide a complete environment to fully influence the coenzyme chemistry. Therefore, the challenge has been to develop viable architectures to allow the detailed examination of coenzyme function and potentially to provide ways in which to exploit the remarkable chemistry of organic coenzymes.

Recent developments in the field of protein design have provided new tools for the study of coenzyme function. In this report we have identified specific coenzymes that have been the subject of investigation from a design perspective. A number of different strategies have been employed to integrate the coenzyme function into a peptide or protein scaffold. These include the chemical modification of well-characterized native proteins. For example the research groups of Kaiser, Suckling and Distefano have prepared and characterized discretely modified derivatives of papain and the fatty acid binding proteins as model systems for flavin-, thiazolium-, and pyridoxamine-dependent transformations. A more recent development has involved the use of synthetic, designed scaffolds together with chemical modification. For example, helix-turn-helix and four-helix bundle motifs have been exploited by Baltzer and Nishino. A possible advantage of such systems is that they can be accessed synthetically and therefore may offer greater opportunities for structure/function analysis through multiple peptide synthesis methods. Another approach that also has this advantage is the building block, or coenzyme amino acid chimera, approach that is now employed in a number of research groups. This approach uses stereoselective synthetic methods to prepare suitably protected amino acid building blocks that integrate coenzyme functionality into the residue side chain. In this way the reactive functionality can be delivered virtually to any location within a polypeptide sequence.

Significant advances have been made in the preparation of discrete macromolecules that include both coenzyme function and a defined polypeptide or protein architecture. Preliminary, but promising, functional studies have been carried out and assay methods developed. While in many cases rather modest effects have been observed, what is significant is that the methodology exists to prepare, characterize, and study defined macromolecular constructs. With new information becoming available on coenzyme-dependent protein catalysts from structural biology and mechanistic enzymology, it should be possible to more fully exploit the remarkable breadth of coenzyme reactivity in tailored synthetic systems.

7
References

1. Breslow R, Dong SD (1998) Chem Rev 98:1997
2. Kirby AJ (1996) Angew Chem Int Ed, Engl 35:707
3. Struthers MD, Cheng RP, Imperiali B (1996) Science 271:342
4. Lovejoy B, Choe S, Cascio D, Mcrorie DK, Degrado WF, Eisenberg D (1993) Science 259:1288
5. Brive L, Dolphin GT, Baltzer L (1997) J Am Chem Soc 119:8598
6. Kortemme T, Ramirezalvarado M, Serrano L (1998) Science 281:253
7. Leussing DL (1986) Model reactions. In: Dolphin D, Poulson R, Avramovic O (eds) Vitamin B6 pyridoxal phosphate: chemical, biomedical, and medical aspects, Part A. Wiley, New York, p 69
8. Dunathan HC, Davis L, Gilmer Kury P, Kaplan M (1968) Biochemistry 7:4532
9. Hayashi H (1995) J Biochem 118:463
10. Okamoto A, Higuchi R, Hirotsu K, Kuramitsu S, Kagamiyama H (1994) J Biochem 116:95
11. Kirsch JF, Toney MD (1993) Biochemistry 32:1471
12. Shaw JP, Petsko GA, Ringe D (1997) Biochemistry 36:1329
13. Albert M, Kjellstrand M, Broo K, Nilsson A, Baltzer L (1998) J Chem Soc Chem Commun 15:1547
14. Kuang H, Brown ML, Davies RR, Young EC, Distefano MD (1996) J Am Chem Soc 118:10702
15. Kuang H, Davies RR, Distefano MD (1997) Bioorg Med Chem Lett 7:2055
16. Kuang H, Distefano MD (1998) J Am Chem Soc 120:1072
17. Ory JJ, Mazhary A, Kuang H, Davies RR, Distefano MD, Banaszak LJ (1998) Protein Eng 11:253
18. Qi D, Kuang H, Distefano MD (1998) Bioorg Med Chem Lett 8:875
19. Banaszak L, Winter N, Xu Z, Bernlohr DA, Cowan S, Jones TA (1994) Adv Protein Chem 45:89
20. Kaiser ET (1989) Acc Chem Res 22:47
21. Buelt MK, Xu Z, Banaszak LJ (1992) Biochemistry 31:3493
22. Sinha Roy R, Imperiali B (1994) J Am Chem Soc 116:12083
23. Sinha Roy R, Imperiali B (1995) J Org Chem 60:1891
24. Sinha Roy R, Imperiali B (1996) Tetrahedron Lett 37:2129
25. Sinha Roy R, Imperiali B (1997) Protein Eng 10:691
26. Imperiali B, Fisher SL, Moats RA, Prins TJ (1992) J Am Chem Soc 114:3182

27. Hofmann K, Frances FM, Limetti M, Montibeller J, Zanetti G (1966) J Am Chem Soc 88: 3633
28. Shogren-Knaak MA, Imperiali B (1999) Bioorg Med Chem in press
29. Struthers MD, Cheng RP, Imperiali B (1996) J Am Chem Soc 118:3073
30. Struthers M, Ottesen JJ, Imperiali B (1998) Folding Design 3:95
31. Shogren-Knaak MA, Imperiali B (1998) Tetrahedron Lett 39:8241–8244
32. Martell AE (1989) Acc Chem Res 22:115
33. Ugai T, Tanaka S, Dowaka S (1943) Yakugaku Zasshi 63:296
34. Yount RG, Metzler, DE (1959) J Biol Chem 234:738
35. Breslow R (1958) J Am Chem Soc 80:3719
36. Lindqvist Y, Schneider G, Ermler U, Sundstrom M (1992) EMBO Journal 11:2373
37. Dyda F, Furey W, Swaminathan S, Sax M, Farrenkopf B, Jordan F (1993) Biochemistry 32:6165
38. König S, Schellenberger A, Neef H, Schneider G (1994) J Biol Chem 269:10879
39. Arjunan P, Umland T, Dyda F, Swaminathan S, Furey W, Sax M, Farrenkopf B, Gao Y, Zhang D, Jordan F (1996) J Mol Biol 256:590
40. Nilsson U, Meshalkina L, Lindqvist Y, Schneider G (1997) J Biol Chem 272:1864
41. Crosby J, Lienhard GE (1970) J Am Chem Soc 92:5707
42. Lobell M, Crout DHG (1996) J Chem Soc Perkin Trans 1 13:1577
43. Washabaugh MW, Jencks WP (1988) Biochemistry 27:5044
44. Golbik R, Neef H, Hübner G, König S, Seliger B, Meshalkina L, Kochetov GA, Schellenberger A (1991) Bioorg Chem 19:10
45. Kern D, Kern G, Neef H, Tittmann K, Killenberg-Jabs M, Wikner C, Schneider G, Hübner G (1997) Science 275:67
46. Tam-Chang S, Jimenez L, Diederich F (1993) Helv Chim Acta 76:2616
47. Hilvert D, Breslow R (1984) Bioorganic Chemistry 12:206
48. Yamashita K, Sasaki S, Osaki T, Nango M, Tsuda K (1995) Tetrahedron Lett 36:4817
49. Aitken DJ, Alijah R, Onyiriuaka SO, Suckling CJ, Wood HCS, Zhu L (1993) J Chem Soc Perkin Trans 1 5:597
50. Suckling CJ, Zhu L (1993) Bioorg Med Chem Lett 3:531
51. Cunningham BC, Wells JA (1997) Curr Opin Struct Biol 7:457
52. Imperiali B, Sinha Roy R, K. WG, Wang L (1996) In: Wilcox CS, Hamilton AD (eds) Molecular design and bioorganic catalysis. Kluwer, The Netherlands, p 35
53. McDonnell K, Imperiali B (1998) submitted for publication
54. Ho SP, DeGrado WF (1987) J Am Chem Soc 109:6751
55. Ghisla S, Massey V (1989) Eur J Biochem 181:1
56. Walpole CSJ, Wrigglesworth R (1987) Oxido-reductases – flavoenzymes. In: Page MI, Williams A (eds) Enzyme mechanisms. The Royal Society of Chemistry, London, p 506
57. Thorpe C, Kim JP (1995) FASEB J 9:718
58. Garrett RH, Grisham CM (1995) Nicotinic acid and the nicotinamide coenzymes. In: Biochemistry. Saunders, Orlando, p 468
59. Shinkai S (1990) Chemical models of selected coenzyme catalysis. In: Suckling CJ (ed) Enzyme chemistry: impact and applications. Chapman and Hall, London, p 50
60. Levine HL, Nakagawa Y, Kaiser ET (1977) Bioch Biophys Res Commun 76:64
61. Rokita SE, Kaiser ET (1986) J Am Chem Soc 108:4984
62. Hilvert D, Kaiser ET (1985) J Am Chem Soc 107:5805
63. Kokubo T, Sassa S, Kaiser ET (1987) J Am Chem Soc 109:606
64. Kaiser ET, Levine HL (1978) J Am Chem Soc 100:7670
65. Slama JT, Radziejewski C, Oruganti S, Kaiser ET (1984) J Am Chem Soc 106:6778
66. Mihara H, Tomizaki K, Nishino N, Fujimoto T (1993) Chem Lett 9:1533
67. Nishino N, Tsunekawa Y, Arai T, Fujimoto T (1995) Synthesis of a catalytic molten globule with flavin function. In: Nishi N (ed) Peptide chemistry. Protein Research Foundation, Osaka, p 485
68. Carell T, Butenandt J (1997) Angew Chem Int Ed Engl 36:1461

69. Oppenheimer NJ, Handlon AL (1992) Mechanism of NAD-dependent enzymes. In: Sigman DS (ed) The enzymes, Academic Press, San Diego, p 453
70. Chenault HK, Whitesides GM (1987) App Biochem Biotech 14:147
71. Adams M (1987) Oxido-reductases – pyridine nucleotide-dependent enzymes. In: Page MI, Wlliams A (eds) Enzyme mechanisms, The Royal Society of Chemistry, London, p 477
72. Zhen J, Chen YQ, Callender R (1993) Eur J Biochem 215:9
73. Anderson L, Suckling CI, Zhu L (1994) Bioorg Med Chem Lett 4:1415
74. Walkup G, Imperiali B (1994) unpublished results
75. Kjellstrand M, Broo K, Andersson L, Farre C, Nilsson A, Baltzer L (1997) J Chem Soc Perkin Trans 2 12:2745
76. Broo KS, Brive L, Ahlberg P, Baltzer L (1997) J Am Chem Soc 119:11362

Functionalization and Properties of Designed Folded Polypeptides

Lars Baltzer

Department of Chemistry, Göteborg University, S-412 96 Göteborg, Sweden.
E-mail: lars.baltzer@oc.chalmers.se

Designed, folded and functionalized polypeptides and proteins constitute an enormous pool of new shapes, new functions and new materials. By taking advantage in the chemical laboratory of the principles of protein folding used by nature, strategies have so far been developed for the engineering of new catalysts, metalloproteins, heme proteins, glycoproteins, receptors and mimics of the components of the immune system. Catalysts have been developed that catalyze reactions not performed by nature and uncommon folded polypeptide motifs have been engineered and structurally characterized. The search for and exploitation of the tremendous number of proteins yet to be discovered has thus begun. Understanding of the protein folding problem has now reached a level where the design of peptides that approach a hundred residues in size is feasible, although not trivial, and clearly sequence dependent. The most frequently designed motif is the four-helix bundle, but recently monomeric triple-stranded β-sheet structures have also been reported as well as a $\beta\beta\alpha$-motif, helical coiled coils and triple helices. Template-assembled polypeptides as well as linear sequences have been shown to fold into designed solution structures and these and other motifs are now key targets for functionalization. This review describes the principles and strategies used in the design of these motifs, as well as their structural characterization. Strategies for functionalization using both the naturally occurring amino acids and post-synthetic incorporation of non-natural functionality will be described, as well as the level of function that has been achieved by rational design.

Keywords: Design, Polypeptide, Catalysis, Metalloprotein, Heme, Structure, Protein folding, Glycopeptide.

1
Introduction

The number of naturally occurring proteins is few in comparison with those that can be made from the twenty commonly occurring amino acids. There are 10^{200} possible 150-residue proteins and there are 10^{38} that have less than 20% sequence homology and that therefore represent unique folds. If the conservative assumption is made that one in a billion of the unique folds will form stable tertiary structures there are still 10^{29} new proteins to be discovered! The number of naturally occurring ones are less than 10^6, probably because they have evolved for the sole purpose of survival. The engineering of tailor-made proteins for new and well-defined purposes therefore presents a considerable challenge and also a tremendous opportunity to chemists. Recent developments in both the understanding of the problems associated with protein folding and in spectroscopic techniques suggest that such endeavors are now timely. The foundations for what may turn out to be well-founded excitement have been laid over the last decade in the many laboratories where the now generally accepted rules for protein design have been elucidated and are due to a large extent to the developments in biomolecular NMR spectroscopy.

The exploitation of the principles of protein folding in the engineering of new proteins is potentially a tremendously rich source of new shapes, new functions and new materials with a degree of complexity that cannot be accomplished by traditional synthesis of organic molecules using comparable experimental effort. In addition, *de novo* designed proteins can be used to study naturally occurring phenomena without the obscuring multiple functions that native proteins perform. Provided that the folded structure of a given amino acid sequence can be predicted, functional groups can be organized in three-dimensional space, in principle at will, to form cooperative sites for a variety of functions. Understanding of the protein-folding problem has now reached a level where non-natural proteins that approach 100 residues in size can be successfully designed. The subtle but important distinction on a molecular level of whether these proteins show the properties of native proteins or have the hallmarks of so called molten globules has been successfully made for a number of motifs and some understanding of how to control it is now emerging [1 – 10].

The engineering of four-helix bundle proteins has been developed in little more than a decade and an increased understanding of how to introduce enough binding energy to make polypeptides fold, as well as to some extent of how to control the formation of well-defined tertiary structures, has now emerged. A high-resolution NMR structure that shows the organization of the residues that make up the hydrophobic core of a *de novo* designed four-helix bundle protein is available, from which it is possible to extract basic design principles [1]. The problems inherent in the engineering of β-sheet proteins haunted the design of the second major fold, while the design of helical motifs was explored because the propensity for hydrogen-bond formation to flanking β-strands favors polymerization and leads to poor solubility. Very recently, however, monomeric triple-stranded β-sheets have been reported in what is a major development in *de novo* protein design [9 – 10] and a high-resolution NMR structure is available that shows the organization of the residues that are crucial to the folding of the motif [9]. The design of mixed motifs with well-defined tertiary structures is timely and a $\beta\beta\alpha$-motif was successfully designed that contained only 23 residues [5]. This peptide set new standards for the number of residues needed to form folded structures and, again, a high-resolution NMR structure was available that described the interplay between the residues in the hydrophobic core.

A research area in which obvious applications are to be found include the design of new catalysts and a number of examples have now been reported of the catalysis of chemical reactions by designed folded polypeptides [11 – 13]. So far, enzyme-like selectivities and efficiencies have not been achieved but, eventually of course, they will be, as strategies for selection for catalysis are developed. What we now see are the early stages of development towards the understanding of how to engineer polypeptide catalysts with tailor-made specificities, with the long-term goal of catalyzing reactions for which no natural enzymes exist. It will be shown in this review that in some cases highly complex reactive sites have been accomplished by rational design and that a key problem is the optimization of geometries in the reactive sites. One of the major discoveries so far is that the binding energies on the surface of folded proteins are strong enough to lead to dissociation constants for protein–substrate complexes in the micromolar range!

Another area of great current interest in protein and polypeptide design is the engineering of metal-binding sites because of the important functions perform-ed in nature by metalloproteins. The incorporation of a metal ion presents special problems because the binding of ligands by metals imposes strong con-straints on the structure of the folded protein. Therefore, while the design-ed ligands clearly bind the metal, as shown by the ensuing UV spectroscopic changes, the three-dimensional structure of the protein may be strongly af-fected. Nevertheless, several reports of designed metal-binding proteins have appeared [14–18] and the incorporation of metalloporphyrins especially has reached a level of complexity that allows detailed biophysical studies that are relevant to the function of their natural role models [19–23]. Future applications of designed heme-binding proteins include the studies of electron-transfer reactions, oxidation–reduction reactions and the engineering of selective and efficient catalysts.

Strategies for functionalization have been developed that exploit the natural-ly occurring amino acids as well as the non-natural ones. Post-synthetic modi-fications have been reported that are based on reactive sites that self catalyze the incorporation of the new functionality at the side chains of Lys residues [24, 25] and on the chemoselective ligation reaction [26–29]. These developments in combination with new methodology for the synthesis of large proteins [30] provide access to a highly versatile pool of new polypeptides and proteins.

As an alternative approach to the design of single-chain proteins that fold into predetermined structures, template-assembled synthetic proteins (TASP) [31–35] have been synthesized, most recently for the formation of four-helix bundle proteins [20] and triple-helix collagen mimics [8]. The TASP approach avoids a central problem in tertiary structure formation as folding is initiated by the template that controls the directionality and proximity of the secondary structure elements. The TASP approach also has a large potential in providing a combinatorial strategy as each secondary structure element can be synthesized individually and a large number of helices can be combined on the template using orthogonal strategies to form a library of proteins. The molecular biolog-ical approach to the problem of generating protein libraries by design has also been demonstrated [36] and been shown to generate folded proteins and folded proteins that bind heme groups [21].

This review describes designed and folded helical proteins, β-sheet proteins, a $\beta\beta\alpha$-motif and TASP proteins that are targets for functionalization. The func-tionalization of folded polypeptides using natural amino acids to form catalysts and heme- and metal-binding proteins will be described as well as post-synthetic functionalization of folded polypeptides to form glycopeptides.

2
Protein Design

In the amino acid sequence of a folded polypeptide is encoded its three-dimensional structure, its folding pathway and the spatial organization of the residues that are responsible for its function (Fig. 1).

The three-dimensional structures of folded polypeptides are controlled mainly by non-covalent interactions and, provided that the folding pathways can be controlled, highly complex molecules can be synthesized by strategies that are by now well developed. The function of the molecule can therefore be the primary goal rather than its synthesis. The understanding of the relationship between the amino acid sequence and the fold and the function provides a tool whereby it will be possible for chemists to design novel molecules with the properties of native proteins for functions that are not performed by nature. A particularly interesting example is the catalysis of chemical reactions, but it may also be possible to imitate native proteins with a role in the life processes and engineer tailor-made mimics where some functions have been incorporated but

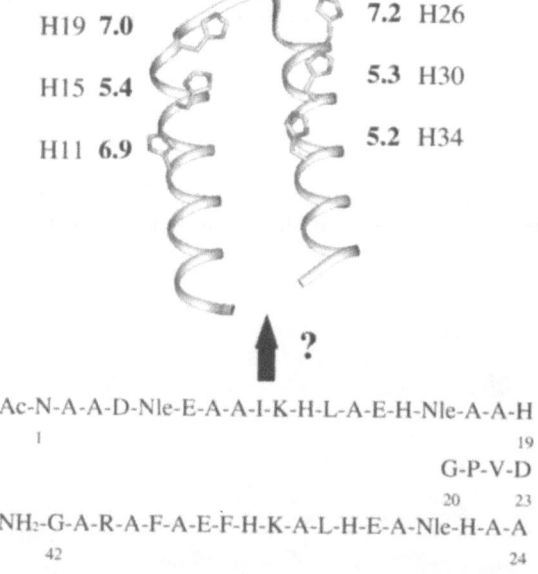

H19 7.0   7.2 H26

H15 5.4   5.3 H30

H11 6.9   5.2 H34

?

Ac-N-A-A-D-Nle-E-A-A-I-K-H-L-A-E-H-Nle-A-A-H
1   19
                                    G-P-V-D
                                    20   23
NH₂-G-A-R-A-F-A-E-F-H-K-A-L-H-E-A-Nle-H-A-A
42   24

Fig. 1. In the amino acid sequence of KO-42 is encoded its fold and its function as it controls the formation of a hairpin helix-loop-helix motif that dimerizes to form a four-helix bundle. On the surface of the folded motif a reactive site is formed that catalyzes hydrolysis, transesterification and amidation reactions of reactive esters, whereas unfolded peptides are incapable of cooperative catalysis. In addition the pK_a values, and thus the reactivities, of the histidine residues are controlled by the fold. The pK_a of each His residue of KO-42 is shown in the figure and deviate by as much as 1.2 units from that of random coil peptides which is 6.4

others have been excluded. In an ideal situation it may be possible, for example, to up and down regulate the response of the immune system.

The rational design approach for linear sequences most often begins with the construction of the secondary structures and when shape complementarity has been introduced so that the super-secondary structures can form, loops or turns are incorporated. In the TASP approach the strategy is mainly to design shape complementary helices and introduce helix to template linkage sites. In both approaches the design of secondary structures is the key issue, with regards to stability and shape, although the highest possible stability in terms of, for example, the free energy of unfolding is not the primary goal in the design of well-defined tertiary structures.

2.1
Design of Secondary Structures

The building blocks in the design of folded polypeptides and proteins are the helix and the β-sheet (Fig. 2). The rational design process starts from the design of these structural elements to provide enough stabilization so that the secondary structures are formed. These secondary structures are not required, however, and are not intended to fold as monomers as most of the driving force for cooperative folding is derived from interactions with other secondary structures. The secondary structure elements are then linked by turns and loops. An excellent review of the design of secondary structures has been published by Bryson et al. [37].

2.1.1
Helices

The design of helical peptides has been the subject of numerous reviews and will not be described in detail here. In short, the helical propensity of the peptide

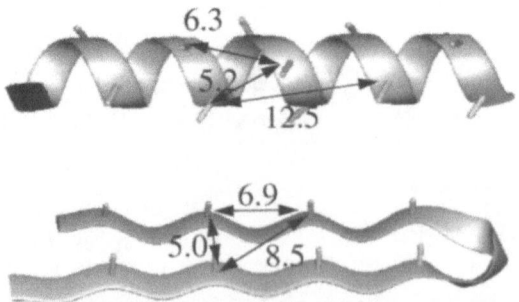

Fig. 2. Schematic representations of the common building blocks of folded polypeptides, the α-helix and the β-sheet. Typical distances between α-carbons in the folded secondary structures are shown and define the dimensions that can be used for the construction of reactive sites in single motifs. The distances between α-carbons of helices depend on the type of helix, and those of β-sheets are very different on the concave and on the convex sides and can easily vary by an Ångstrom or more.

depends on the helical propensities of the amino acid residues and the folded helical segment can be further stabilized by a number of inter-residue interactions with typical stabilizing energies [37]. The difference in helix propensity between individual residues can amount to 1 kcal/mol. Neutralization of the N- and C-terminal charges by acetylation and amidation, respectively, prevents repulsion by the helical dipole, and introduction of charged residues of opposite signs close to the termini stabilizes the helix dipole (0.5 kcal/mol). A judicious choice of N- and C-terminal residues also provides improved capacities for hydrogen bonding to amide and carbonyl groups in the first and last turns of the helix (N-cap 1–2 kcal/mol, C-cap 0.5 kcal/mol). In addition, salt bridges may be introduced on the surface of the folded helix, e.g. by incorporating lysine and glutamic acid residues four positions apart in the sequence (0.5 kcal/mol). However, in folded tertiary structures the binding energies between hydrophobic residues dominate the folding of the peptide and these energies are in the order of 2 to 5 kcal/mol per hydrophobic residue and in terms of the formation of secondary structures in folded motifs the long-range interactions are the most important. The use of a helix as a structural template for functionalization provides a molecular scaffold where the inter-residue distances are well defined.

2.1.2
Sheets

The rules for β-sheet formation are much less well understood than those for α-helix formation. However, β-sheet propensity scales have been reported [38, 39] and inter-residue stabilization by salt-bridge formation and hydrophobic interactions provide extra binding energy in the folded state of a similar magnitude to that in helices. The β-sheet is by definition long range in origin and although inter-residue interactions are important in controlling topology well-defined rules that govern the state of aggregation remain to be elucidated. As in the case of the helix the β-sheet provides a well-defined molecular scaffold where the inter-residue distances are well understood. So far, however, there have been no reports of de novo designed functionalized β-sheets.

2.2
Design of Supersecondary Structures

Secondary structures may be combined to form supersecondary or tertiary structures if enough binding energy is available to drive the formation of the folded motif, and the designed structure is likely to form if shape complementarity can be introduced to control the docking of the secondary structures. The most common designed motif is that of the four-helix bundle formed from amphiphilic helices where shape complimentarity and hydrophobic interactions drive the formation of the tertiary structure [37]. The size of the helical segments has been found to be critical and as a rule of thumb approximately 15–20 residues are needed to fold, giving rise to four-helix bundle motifs that contain 60–80 residues. A common motif is the helix-loop-helix dimer formed from sequences of 35–43 residues [1,2] that fold and dimerize in solution. The size of designed motifs can, however, be considerably smaller than that, as has been shown for the β-sheet Betanova and the $\beta\beta\alpha$-motif BBA1 that contain 20 [9] and 23 residues [5], respectively.

2.2.1
Amphiphilic Helices

The design of amphiphilic helices is based on the repetitive nature of helical segments. In the coiled-coil motif a heptad repeat, $(abcdefg)_n$, is found where the a and d positions are populated by hydrophobic residues (Fig. 3). This heptad repeat pattern has been successfully used in the design of coiled-coil proteins [6] but less regular patterns also give rise to ordered folded structures [2]. Several designed polypeptides that have been reported to fold into highly helical structures have little regularity except that at least every fourth residue in each helix is hydrophobic [37]. This periodicity ensures that one face of each helix is hydrophobic and if approximately one hydrophobic residue per turn of each helix is introduced enough binding energy is available to drive the folding with typical dissociation constants for the dimer–monomer equilibrium in the µM range. In the main, these motifs have been characterized by CD spectroscopy because ^{1}H NMR investigations are hampered by broadened resonances and poor chemical shift dispersion. Consequently, little is known about the organization of the hydrophobic core, except that it is disorganized!

2.2.2
The Turn-Loop Problem

The connection of secondary structures in native proteins is controlled by ordered structures, turns, or unordered structures, loops. In protein design the

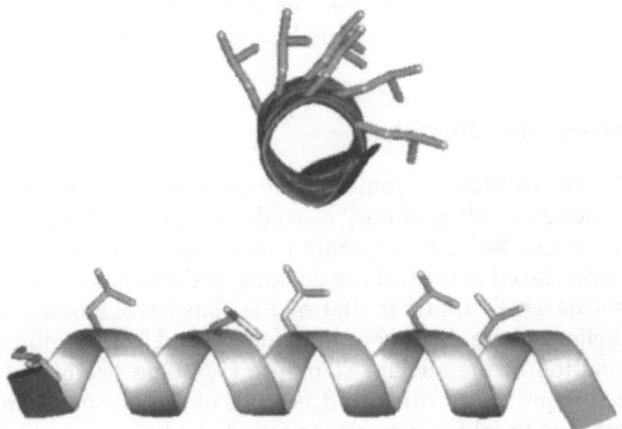

Fig. 3. Schematic representation of the organization of hydrophobic residues on the surface of a folded helix based on the heptad repeat, $(abcdefg)_n$, where a and d are hydrophobic

design of turns most often leads to the construction of loops! This difficulty emanates from an incomplete understanding of how to terminate helices and β-strands so that the exact location of the N- and C-terminal amide groups is not known with a high degree of accuracy. In order to circumvent this problem rigid templates and β-turn mimics have been synthesized and used with a large degree of success, e.g. the Kemp triacid [40] or small cyclic peptides [32, 33]. On the other hand, the dominant contribution to the folding energy of larger motifs comes from the hydrophobic interactions between amphiphilic secondary structures and the connecting loop thus plays a less important role in most cases in controlling the tertiary structure, except for the regulation of the overall fold. For this reason the structure of the loop need not be known at an atomic level although in designed proteins with native-like properties the loop may be crucial to the formation of the tertiary structure [5, 10].

2.3
The TASP Strategy

In order to avoid the problems inherent in controlling the overall fold of a designed protein and to provide routes for combinatorial approaches the idea of template-assembled synthetic proteins (TASP) was introduced by Mutter [31] (Fig. 4). A cyclic peptide was used as the template [32, 33] and the amino acid side chains that were found on one side of the cyclic peptide were used as linkers for the addition of linear peptides. Orthogonal protection group strategies were subsequently developed that allowed the incorporation of several different sequences site selectively [41]. The initiation of folding by the template provides, in theory, a gain in entropy over that of the linear peptides by reducing the number of degrees of freedom of the unfolded state. The problems that are encountered in the design of polypeptides based on the TASP strategy are similar to those that have to be overcome in the design of metal-binding sites as the structural constraints imposed by the scaffold itself are very demanding. A partial solution to this problem is to make use of "spacers", for example, a sequence of glycine residues, flexible fragments that allow the secondary structures to form the best interfaces possible. So far, few of the peptides that have been designed according to the TASP approach have been structurally characterized by detailed NMR spectroscopic analysis, probably because of the poor chemical shift dispersion and extensive line broadening that arises when the folded polypeptide is in equilibrium between several conformations. The exception is the collagen mimic triple-helix motif based on the Kemp triacid, which has been structurally characterized by NMR spectroscopy and found to have a well-defined tertiary structure [8].

An advantage of the TASP approach is that it provides an efficient combinatorial strategy for protein engineering because the orthogonal coupling strategies allow the controlled introduction of different secondary structures. The efficient synthesis of e.g. 20-residue helical segments of high purity in large numbers by automated synthesizers makes this strategy highly suitable for the development of large libraries of functionalized TASP proteins.

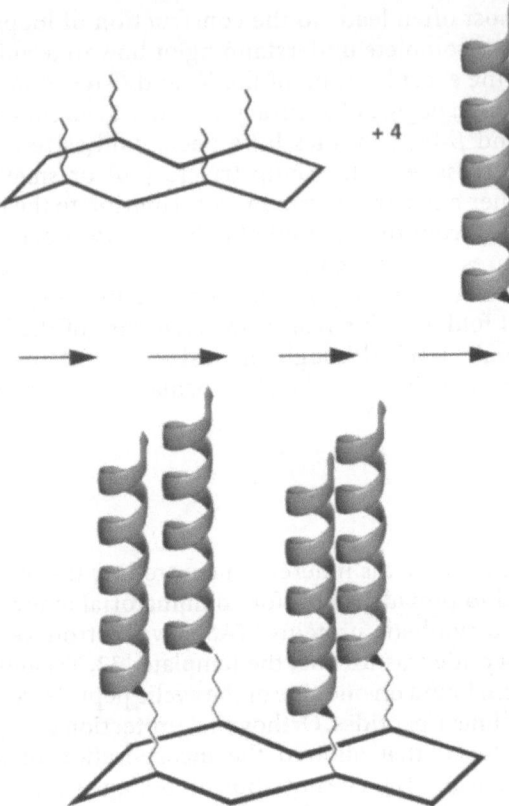

Fig. 4. Schematic representation of template-assembled synthetic proteins. The conformationally restricted template can be orthogonally protected and sequentially linked to helical segments to form a large variety of functionalized TASP proteins. Flexible spacers that connect the folded peptide segments and the template provide the necessary conformational freedom that will allow the hydrophobic residues to find their optimum orientations for packing the core

2.4
The Molecular Biological Approach to Protein Design

The machinery of the living cell provides in many contexts a superior way of producing biomolecules although the limitations of the selection techniques still constitute a formidable problem in the extraction of the optimum species from a large pool of proteins. In an attempt to employ such a strategy in the engineering of four-helix bundle proteins with well-defined tertiary structures, the hydrophobic residues in the heptad repeat of a linear sequence designed to form the hydrophobic core of the folded motif were varied [36]. While the selection of folded four-helix bundle proteins still creates difficulties, proteins have been selected from the generated library that show impressive properties [4, 21].

2.5
What Makes Tertiary Structures Well-Defined?

Much of the interest in *de novo* protein design stems from an incomplete understanding of the so-called protein-folding problem. The ultimate test of the level at which we can claim to know how and why proteins form tertiary structures is to be able to design a protein from first principles that folds cooperatively into a predetermined and unique structure. This problem is twofold, the first question is what size amino acid sequence is required to provide enough binding energy to drive the folding process. The answer to this question appears to depend on the motif, and peptides with approximately 20 residues have been shown to form monomeric $\beta\beta\alpha$- and β-structures in solution [5, 9] whereas folded helical proteins often approach 80 residues in size, or half of that if it is a dimer of helix-loop-helix motif, etc. The second and perhaps more difficult question is what principles govern the formation of proteins with unique folds.

In helical structures based on amphiphilic segments it appears that approximately every fourth residue of each helix should be hydrophobic and that the extra stabilization provided by inter-residue salt bridges, capping, etc. is of less importance, although it increases the free energy of unfolding by the tabulated amounts. The level of understanding is now such that the probability of success is high in obtaining a helix-loop-helix motif that dimerizes to form a four-helix bundle if the sequence contains approximately 40 residues that include two segments with high helical propensity and a loop sequence that is similar to those that occur in naturally occurring β-turns [37, 42, 43, 45].

The solution to the problem of how to induce native-like properties is more complex and to date only partial solutions exist. It was realized some time ago that maximum hydrophobic binding energy was not the solution to the protein-folding problem in terms of providing a unique fold. In fact, the somewhat surprising lesson may turn out to be that too much binding energy, particularly in stabilizing secondary structures, is detrimental to the cooperative formation of well-defined tertiary structures. The definition of what is a native protein is a difficult one but commonly used criteria include the appearance of a well-dispersed ^{1}H NMR spectrum that is in slow exchange on the NMR timescale (Table 1), well-defined thermal melting points and slow exchange rates of back-

Table 1. Chemical shift dispersion in ppm of the designed and naturally occurring four-helix bundle proteins α2D [1], GTD-43 [2], SA-42 [42] and IL-4 [78] [a]

	NH	αH	Methyl
SA-42 [b]	1.12	0.9	0.15
α2D	2.02	1.20	0.62
GTD-43	2.16	1.35	0.85
IL-4	2.63	1.93	0.92

[a] Values given for amide protons (NH), α-protons (αH) and methyl groups of residues in the hydrophobic core (Methyl).
[b] Assignment in 14 vol% trifluoroethanol.

bone amide protons. The ability to bind hydrophobic dyes is, however, inconclusive since, for example, chymotrypsin binds ANS. So far, four-helix bundle proteins, triple helices, β-sheets, a ββα-motif and a coiled-coil motif have been shown to fulfill some of these criteria [1 – 10].

Helical heptad repeat sequences have been reported to be well behaved although they are difficult to characterize by NMR spectroscopy due to spectral overlap. The motifs that have been shown to have native-like properties, and are not highly repetitive, have cores composed of aromatic amino acid side chains of, for example, phenylalanine and tryptophan. In four-helix bundle motifs [1, 2], the ββα-motif BBA1 [5] and the β-sheet protein Betanova [9], the formation of the folded structure appears to be strongly dependent on such residues although the energetics have not been calculated by substitution studies. As a tentative rule, therefore, the probability of success in the design of a new protein is probably much higher if residues are included that can form aromatic clusters in the core (Fig. 5).

The conclusions to be drawn so far about what is needed in a design to induce a native-like fold are therefore that the sequences should either show a high degree of symmetry and be based on the heptad repeat [6], or they should have a hydrophobic core that is shape complementary and at least partly aromatic [1, 2, 5, 9].

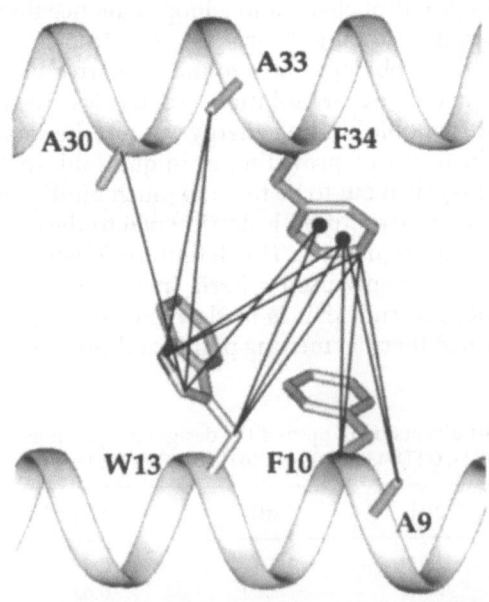

Fig. 5. The aromatic cluster of the hydrophobic core of GTD-43, a helix-loop-helix dimer, and some of the assigned long-range NOEs that demonstrate the interactions of the aromatic side chains in the folded motif. The formation of aromatic clusters has been observed in several designed proteins. Reproduced with permission from J Am Chem Soc (1997) 119:8598. (©1997 ACS)

3
Designed Proteins

A large number of proteins have been designed to date but the purpose of this review is not to list them all. The interested reader is referred to a series of excellent review articles by DeGrado et al. [37, 44]. Here, the focus will rather be on those proteins that are of importance in the development of *de novo* design and those that have been shown to play a role in functionalization to demonstrate concepts and trends.

3.1
Four-Helix Bundle Proteins

The four-helix bundle protein is probably the most common motif in protein design most likely because it is based on the formation of a motif from what is essentially self-contained structural entities. For well over a decade DeGrado et al. have published excellent work on the systematic elucidation of the rules that govern the formation of four-helix bundle proteins with well-defined tertiary structures which has provided many of the insights into the magnitude of binding energies and of design rules for hydrophobic cores [45–47]. The crowning achievement was the publication of an NMR structure at atomic resolution of the polypeptide α_2D, a 35-residue polypeptide that folds into a helix-loop-helix motif and dimerizes to form a four-helix bundle protein [1] (Fig. 6).

The formation of aromatic clusters from Phe and Trp in the hydrophobic core of the folded motif was clearly shown and these were identified as possible contributors to the native-like properties of the designed protein. The ^1H NMR spectrum was well dispersed with narrow line widths in agreement with what is expected from a native-like protein (Table 1). These properties have been gradually achieved in a series of helix-loop-helix motifs in a rational design process. α_2D shows a thermodynamic stability that is sufficient to allow substantial modification and should therefore be an excellent scaffold for functionalization, in particular since deep insights have been accumulated as to how to reengineer the protein towards a more stable conformation if the need should arise as the sequence is changed. The structure of the peptide was somewhat surprising in that it formed an antiparallel interleaved dimer rather than the most frequently modelled antiparallel dimer where the two hairpin motifs are just added together. The formed motif is unusual in nature and shows that new motifs are accessible in man-made designs.

Similar properties were reported by Dolphin et al. for the designed 43-residue peptide GTD-43, that also folds into a hairpin helix-loop-helix motif and dimerizes [2]. While an atomic resolution structure does not yet exist, extensive NMR spectroscopic investigations, including, for example, 19 long-range NOEs and slow amide proton exchange rates, reveal that here, too, the nucleus of the hydrophobic core is centered around an aromatic cluster that is formed from Phe and Trp residues (Fig. 5). The fact that these two *de novo* designed four-helix bundle proteins both show some properties of native proteins strongly implies that the introduction of aromatic ensembles is a functional design strategy for the

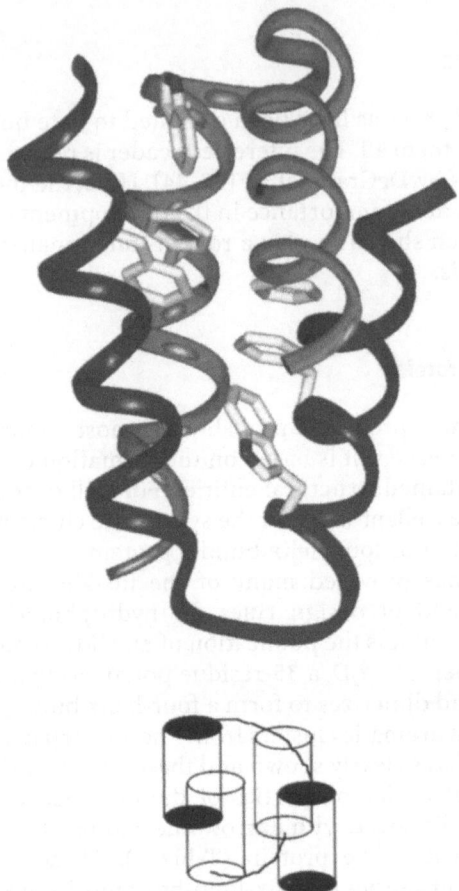

Fig. 6. The high-resolution NMR structure of α_2D, a 35-residue designed polypeptide that forms a four-helix bundle structure, showing the location of the aromatic residues in the core of the folded motif. The cartoon illustrates the unexpected fold of α_2D where the hairpin subunits dimerize in a interleaved mode. Reproduced with permission from J Am Chem Soc (1998) 120:1138. (© 1998 ACS)

engineering of designed native-like proteins. The question of how subtle the balance is between ordered and disordered structures and, consequently, how sensitive the motif is to the introduction of new functions, has not yet been addressed in the case of α_2D, but a modest modification of the GTD-43 motif was successfully undertaken in the peptide GTD-C [48], where a functionalization site was introduced without disrupting the well-defined structure. In contrast, when a buried salt bridge in the hydrophobic core was replaced, GTD-C turned into a molten globule structure.

An NMR structure determination has also been reported of a helix-loop-helix motif where the fold is controlled by the formation of an intramolecular disulfide bridge. The long-range NOEs were limited to residues in close proximity to the linkage site [49] whereas the structure in more remote parts of the motif was disordered. The introduction of disulfide bridges is therefore most likely not a shortcut to well-defined tertiary structures. A helix-loop-helix motif was also reported together with its NMR spectroscopic structural characterization. The fact that a large number of long-range NOEs were obtained suggests that the structure is ordered. The hydrophobic core is composed of mainly aliphatic side chains and its further analysis may therefore provide information on how to engineer non-aromatic cores [50]. The exploitation of the tools of molecular biology led to the selection of a protein that also shows some of the properties of a native protein, although an NMR structure is not likely to be available due to an unfavorable monomer–dimer equilibrium [4].

Traditionally, the structural characterization of designed proteins is carried out by CD spectroscopy, which unfortunately provides only limited structural information at the atomic level. As the understanding of protein design develops more proteins appear that have well-defined structures and the determination of their solution structures by NMR spectroscopy is clearly the main tool for elucidating structure–function relationships. Key information is obtained simply from the 1D spectrum (Fig. 7).

The chemical shift dispersion (Table 1) and the temperature dependence of the resonance line shape provides a qualitative measure of whether the structure is well ordered [2]. However, NMR spectroscopy also provides information relevant to the problem of protein folding in the study of the molten globule states. NMR spectroscopic investigations of molten globules may be more demanding than those of ordered proteins due to spectral overlap arising from poor shift dispersion and to short relaxation times that are due to conformational exchange at intermediate rates on the NMR time scale.

Significant improvements in the appearance of the ^{1}H NMR spectrum of molten globule states are, however, observed upon addition of small amounts of trifluoroethanol (TFE), less than 10 vol%. Amide proton exchange rates are reduced and resonances are sharpened although the chemical shift dispersion is also decreased [51]. The net result is an increased resolution that simplifies considerably the spectral assignment. The increased spectral resolution is in fact due to the fact that the peptides become more denatured rather than more structured and that, in order to obtain more structural information about the folded state, the peptide is partially unfolded!

Using these techniques structural information about a number of four-helix bundle proteins [13, 42] was obtained in spite of the fact that their hydrophobic cores were partially disordered, the chemical shift dispersion was poor and the resonances were extensively broadened in aqueous solution. The dominant solution conformations were identified in mixtures of water and TFE, with typical TFE contents of 2–6 vol%. Assignments of the ^{1}H NMR spectra were readily obtained and the secondary structures were identified from αH chemical shift indices and medium-range NOEs. The state of aggregation was determined by equilibrium sedimentation ultracentrifugation, and the formation of the hairpin motif and of an antiparallel helix-loop-helix dimer was identified from long-range NOEs [51]. The assignment of the NMR spectrum and the identification of the secondary and tertiary structures made it possible to deter-

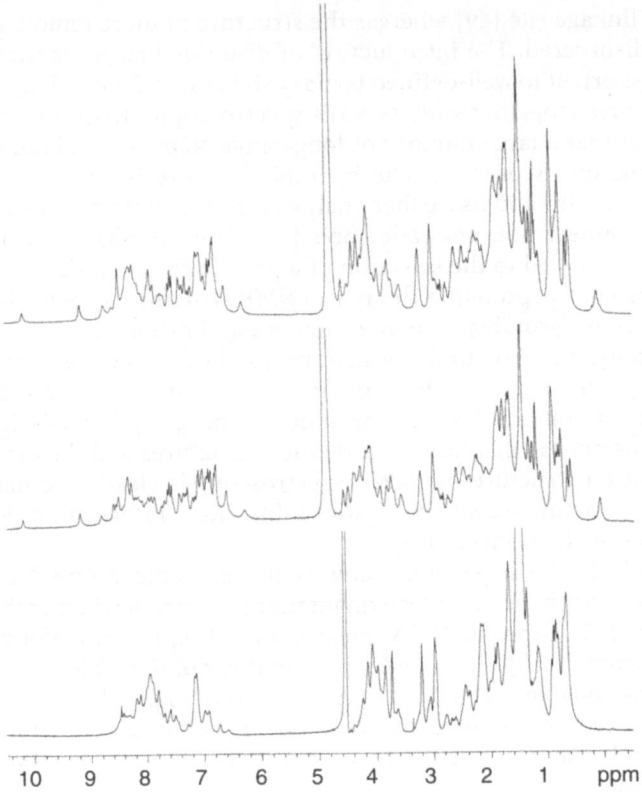

Fig. 7. One-dimensional ¹H NMR spectra of the designed four-helix bundles SA-42 (*lower trace*) and GTD-43 (*top two traces*). The chemical shift dispersion of SA-42 in 90% H_2O and 10% D_2O at 323 K and pH 4.5 is poor and the resonances are severely broadened due to conformational exchange. The chemical shift dispersion of GTD-43 in the same solvent at 288 K and pH 3.0 is comparable to that of the naturally occurring four-helix bundle IL-4 and the resonances are not significantly affected by conformational exchange. Upon raising the temperature to 298 K line broadening is observed (*top trace*) which shows that GTD-43 is in slow exchange on the NMR time scale, unlike SA-42 where an increased temperature reduces the line width. These spectra are therefore diagnostic of structures with disordered (SA-42) and ordered (GTD-43) hydrophobic cores

mine the reaction mechanism of the four-helix bundle catalyst KO-42 [13] partly because the pK_a values of six histidine residues were unequivocally assigned. NMR spectroscopic studies can thus provide structural information on many levels and is an indispensable tool in protein design and functionalization.

3.2
β-Sheet Proteins

While many studies of helical polypeptides and proteins were reported and the understanding of how to engineer four-helix bundle proteins increased the design of β-sheet proteins was notoriously difficult. The designed peptides had a strong tendency to aggregate and precipitate due to the inherent propensity of β-structures for hydrogen bonding to other β-strands. Solubility was consequently low and the state of aggregation ill defined. It is only recently that designed, well-defined and monomeric β-structures have been reported, and the engineering of the 20-residue peptide Betanova that folds into a triple-stranded β-sheet and is monomeric over a wide range of concentrations is an achievement in protein design [9] (Fig. 8).

In this structure, no disulfide bridges are used, only naturally occurring amino acids. Its NMR structure shows the diagnostic long- and medium-range NOEs typical of β-folds and several specific interactions were identified, in particular, the packing of a tryptophan residue in a hydrophobic cluster. Further studies of Betanova may contribute to an understanding of how to control the state of aggregation in designed β-sheets. A monomeric triple-stranded β-sheet protein was also reported where a d-proline was used as well as a non-natural amino acid [10]. These advances provide the opportunity for functionalization of β-structures and perhaps also an answer to the question of how extensive modifications can be undertaken without disrupting relatively small motifs.

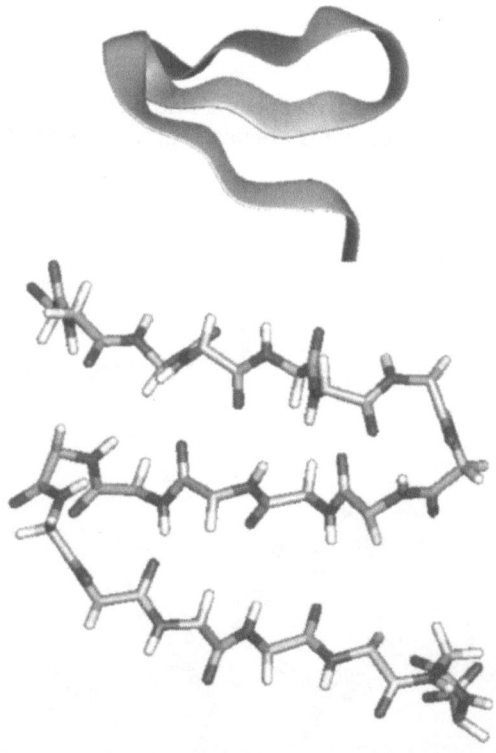

Fig. 8. Peptide backbone structure of the designed triple-stranded monomeric β-sheet Betanova in band and liquorice model representation, coordinates taken from high-resolution NMR structure [9]

3.3
The ββα-Motif

Mixed motifs are probably necessary for optimal functionalization since they provide greater versatility in design due to larger variation in site geometries. It is likely that the most successful catalysts will turn out to be the ones that have reactive site residues in separate secondary structures. The impressive design of a 23-residue peptide, BBA1, that was shown to fold into a ββα-motif, therefore, provides the possibility of developing complex sites in minimal proteins [5, 52] (Fig. 9). It is based on the structure and sequence of the zinc-finger motif but was redesigned to fold in the absence of Zn! An atomic resolution NMR structure showed key contacts between hydrophobic residues in a core that was centered around a non-natural amino acid, 3-(1,10-phenantrol-2-yl)-l-alanine, but this residue has in more recent designs been replaced by naturally occurring residues, without loss of structural uniqueness [53]. Some indication has been obtained that small motifs are vulnerable to amino acid substitutions, although there is hope that they can be repacked to form well-defined tertiary structures after functionalization. The demonstration that this motif can be successfully redesigned suggests that it can also be successfully functionalized.

3.4
Coiled Coils

The coiled-coil motif is not strictly a specific fold but the description is applied to structures where helical segments are based on the heptad repeat $(abcdefg)_n$.

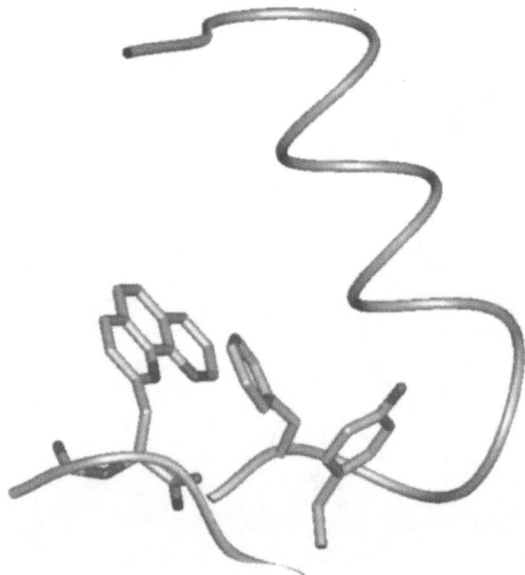

Fig. 9. Structure of BBA1 a ββα-motif showing the fold and the aromatic residues that form part of the hydrophobic core, coordinates taken from high-resolution NMR structure [79, 80]

It is observed in four-helix bundles, in triple helices and in helix dimers of high symmetry. Due to the use of repetitive sequences, NMR spectroscopic investigations are difficult because of severe overlap, although functionalized peptides are expected to be more readily analyzed since the symmetry is partly disrupted. The structural characterization of peptides based on repetitive sequences is most often based on the analysis of other spectroscopic techniques such as the temperature and concentration dependence of the CD spectrum. Native-like properties have been observed in a four-helix bundle coiled-coil motif [6]. The functionalization of dimeric coiled coils has been carried out extensively by Hodges and co-workers [54] and the motif has been found to be very tolerant towards modifications of the amino acid sequence, probably because there is a large number of hydrophobic residues in the core that provide large amounts of binding energy. It is therefore an excellent template for functionalization provided that the geometries of the populated rotamers of the residue side chains fit the target function of the functionalized peptide.

3.5
Four-Helix Bundle TASP Molecules

TASP molecules circumvent the problem of secondary structure orientation by controlling directionality and state of aggregation. The concept of using a structural template for this purpose was pioneered by Mutter more than a decade ago [31] but so far applications have been limited in spite of its appeal. A renewed interest appears to have been initiated [20, 55] with the demonstration of the combination of the TASP strategy to what may be a developing combinatorial approach [55]. The synthesis of 20-residue helical segments are by today's standards straightforward and can be automated. In combination with orthogonal protection strategies large numbers of four-helix bundle proteins can be generated in a short time (Fig. 4). Provided that it is possible to select for function, the TASP strategy may well provide an efficient route towards functionalized designed proteins.

One problem that must be addressed is that of structure. Few NMR spectroscopic measurements have so far been reported of designed TASP molecules and the difficulties here are the same as those of other designed proteins as it is at the present level of protein design very difficult to introduce several structural constraints simultaneously. Covalent bonds between helices and template are well defined with regards to lengths and angles and will allow the formation of well-defined tertiary structures only if the hydrophobic residues that make up the core of the folded protein can be introduced in close proximity to optimal positions. The functionalization will of course further complicate the design and it will be a major accomplishment in protein design if combinatorial approaches can be used to overcome these fundamental problems. It appears that the use of "loose" structural constraints such as glycine spacers alleviate the covalent constraints imposed by the template [8].

3.6
Triple Helices – Collagen Mimics

Although the designed collagen mimics based on the Kemp triacid [40] are not usually referred to as TASP molecules the strategy is the same in that folding is initiated and directionality is controlled by the template. The collagen mimics reported by Goodman et al. [8, 56–59] have been extensively characterized by NMR spectroscopy as well as by other techniques and show excellent structural behavior in that the conformations are constrained and the tertiary structure appears to be well defined. The glycine spacers of the triple helices apparently allow the conformational freedom that allows the hydrophobic residues to find the conformation (or small group of similar minima) that has the lowest free energy and exlude the conformations that in the absence of the template would have similar free energies but different structures. So far, the sequences are highly repetitive and it remains to be seen how tolerant this motif is towards modification and thus functionalization. One strength of this motif is that sufficient binding energy can readily be introduced simply by elongation of the helical segments and it is to be expected that functionalization should not be a problem in other ways in that the geometries may not be perfectly compatible with the desired function. This problem is of course common to all designed proteins of this size.

4
Strategies for the Functionalization of Designed Folded Polypeptides

Considerable freedom of choice is available to the protein designer with regards to the nature of the amino acids, the method of production and the size of the peptide. Designed polypeptides composed of only naturally occurring amino acids will probably provide the most accurate knowledge of how native proteins function although artificial amino acids may be very useful in the elucidation of mechanism. For example, whether or not a binding residue is located in the optimum position to bind a substrate can be determined by making its side chain longer or shorter by replacing arginines and incorporating homoarginines, or by replacing lysines with ornithines [62]. There are also several examples of how designed peptides fold better when amino acids of non-natural stereochemistry are incorporated into loops [5, 10]. The TASP approach and chemoselective ligation strategies are based on the use of non-natural amino acid residues.

Sequences with natural residues can be produced by protein engineering or by peptide synthesis whereas sequences with non-natural residues are, in practice, limited to synthesis. For complete isotopic labelling, protein engineering is the only alternative for reasons of cost but specific labels can in practice only be incorporated by synthetic methods. With current developments in protection group strategies and coupling reagents, peptides of 50 residues in size are readily synthesized on automated synthesizers and the ligation of 50-residue fragments is possible provided that the sequence contains Cys residues [30]. Peptide synthesis may in many cases be preferable to the use of bacterial strains, not only where specific labels or non-natural amino acids are required.

The size and complexity of the motif is clearly important, single helices are of limited use because of the long intra-residue distances on the helical surface (Fig. 2) and functions that require more than two-residue sites will most likely have to depend on the functionalization of supersecondary structures. The clever designs of catalysts for decarboxylation and ligation reactions are, on the other hand, good examples of how small motifs can be exploited for complex functions [11, 12].

4.1
Functionalization Using the Common Amino Acids

The functionalization of folded motifs is based on an understanding of secondary and tertiary structures (Fig. 2) and must take into account the relative positions of the residues, their rotamer populations and possible interactions with residues that do not form part of the site. For example, glutamic acid in position *i* has a strong propensity for salt-bridge formation, and thus reduced reactivity, if there is a Lys residue available *i-4* in the sequence, but the probability is much less if the base is *i-3* [60]. Fortunately, there is a wealth of structural information on the structural properties of the common amino acids from studies of natural proteins that provides considerable support for the design of new proteins. The naturally occurring amino acids have so far been used to construct reactive sites for catalysis [11–13], metal- and heme-binding sites [14, 15, 19, 21, 22] and for the site-selective functionalization of folded proteins [24, 25].

4.1.1
The Template Approach

The motifs used so far for functionalization have been helical whereas functionalized β-sheets have not been reported. It is, however, very likely that functionalized sheets will appear now that monomeric β-sheet proteins have been successfully designed [9, 10]. The dimeric helix-loop-helix motif is very tolerant of modification [13, 43, 61], perhaps because this motif has a large amount of intrinsic binding energy. The replacement of less than five residues has been shown not to be detrimental to the overall fold even though the free energy of unfolding may vary by several kcal/mol [43, 48]. In the elucidation of the function of the catalytically active four-helix bundle KO-42 more than 25 different peptides were synthesized [43] in order to systematically vary the structure of the reactive site but the residues that control folding, mainly the hydrophobic residues in the core, were not altered and the overall fold remained unchanged. Structural characterization by NMR and CD spectroscopy and ultracentrifugation was obtained and all of the reported mean residue ellipticities, from $-20\,000-25\,000$ deg cm^2 dmol^{-1}, were in good agreement with the formation of highly helical proteins and thus the designed fold.

4.1.2
The Use of Non-specific Binding To Overcome Limitations of Crude Templates

The demand for preorganization and precise positioning in the construction of functional sites may limit the degree of success that can be accomplished in simple motifs. In a helix the changes incurred by moving a residue one turn along the helix corresponds to a change in its position by 5.2–6.3 Å, a truly dramatic change in position for a catalytic site residue (Fig. 2). Two residues that are one position apart in the sequence of a helix are directed almost at right angles to each other. Fine tuning of a catalytic site is therefore not possible beyond that level of resolution. Although it is expected that in isolated cases, where the geometrical constraints in the transition state happens to coincide with those of the polypeptide catalyst, spectacular results will be accomplished even in designed motifs, it is unlikely that the majority of the designed sites will be capable of enzyme-like catalysis. The exploitation of hydrophobic binding therefore provides an extremely efficient and advantageous strategy in constructing binding and catalytic sites.

Hydrophobic patches in the folded motif allow several modes of binding of hydrophobic substituents and provide substantial amounts of binding energy. In combination with charged residues the use of hydrophobic sites therefore offers a powerful design strategy for the introduction of both binding efficiency and selectivity. A case in point is the binding of hydrophobic and charged peptide substrates by amphiphilic helices with dissociation constants in the µM range and a high degree of selectivity [12].

4.2
Functionalization Using Non-natural Amino Acids

The incorporation of non-natural amino acids is a viable alternative in protein design using standard peptide synthesis techniques as, for example, in the incorporation of protected homoarginine and ornithine residues [62] and in the incorporation of amino acids for chemoselective ligation [26–30]. In contrast, the incorporation of complex and bulky side chains such as those of protected sugars and cofactors may cause low yields in peptide synthesis and high costs due to the cumbersome synthesis of enantiomerically pure and protected reagents. An appealing alternative is the post-synthetic incorporation of functionality into folded proteins that often does not require extensive efforts in the synthesis of the functional group. The TASP proteins has so far proven to be an interesting mix of natural as well as non-natural amino acids. The template is often non-natural as well as the linker residues whereas the functionality is provided by the common amino acids.

4.3
Post-synthetic Functionalization

Post-synthetic functionalization reactions have been reported that make it possible to introduce new groups into folded proteins and peptides using natural [24, 25] as well as non-natural amino acids [26–30]. The disulfide exchange reaction and the nucleophilic substitution reaction at the side chain of cysteine are by now well known and will not be discussed here, but it should be noted that in order for these reactions to be site selective in natural proteins they are limited to the ones that have a single exposed cysteine residue. For cysteine residues in synthesized polypeptides orthogonal protection group strategies are available [55]. Two alternative approaches have been reported that make use of natural as well as of non-natural amino acids, site-selective functionalization and chemoselective ligation.

4.3.1
Site-Selective Functionalization – Using the Common Amino Acids

Histidine residues are efficient nucleophiles in aqueous solution at pH 7, much more so than lysines, and this is the basis for the site-selective functionalization of lysine residues in folded polypeptides and proteins [24, 25]. *p*-Nitrophenyl esters react with His residues in a two-step reaction to form an acyl intermediate under the release of *p*-nitrophenol followed by the reaction of the intermediate with the most potent nucleophile in solution to form the reaction product. In aqueous solution the reaction product is the carboxylic acid since the hydroxide ion is the most efficient nucleophile at pH 7. If there is an alcohol present the reaction product will be an ester and the overall reaction is a transesterification reaction.

The intermediate can, however, also be trapped by an amine to form an amide although at pH 7 in aqueous solution primary amines are predominantly protonated and only poorly reactive. Intramolecularity will, however, improve the poor reactivity of a lysine residue towards an acyl intermediate provided that the His and the Lys residues are close in space. The net reaction under these conditions is therefore an amidation of the lysine side chain by the active ester that is more efficient than the direct acylation of a lysine residue by at least three orders of magnitude (Fig. 10). The lysine residue will also improve the reactivity of the His side chain by electrostatic transition state stabilization and the wasteful reaction with other His residues that gives rise to hydrolysis is therefore suppressed.

Site selectivity has been investigated in helical sequences and it is only when the Lys residue is in position *i-3* or *i + 4* relative to a His in position *i* that it will be amidated and then the latter will be functionalized in preference to the former [63]. It is thus possible to incorporate acyl groups practically at will since the reactivity of His residues depends mostly on the value of pK_a and it is possible to vary the pK_a of His residues by rational design over a range of 2 pK_a units [60], which corresponds to approximately two orders of magnitude in reactivity for nucleophilic reactions. Interestingly, approximately 20% of the naturally occurring proteins contain the His (*i*) Lys (*i + 4*) configuration, although investigations have not been carried out as to whether the sites occur in helical segments. It has also not yet been elucidated whether similar reactive sites are found in β-structures.

Using this reaction fumaryl groups [24], cofactors [64] and carbohydrate derivatives [65] have been incorporated site selectively into folded four-helix bundle proteins. The reaction can also be expected to be of general use in site-selective immobilization reactions of folded proteins.

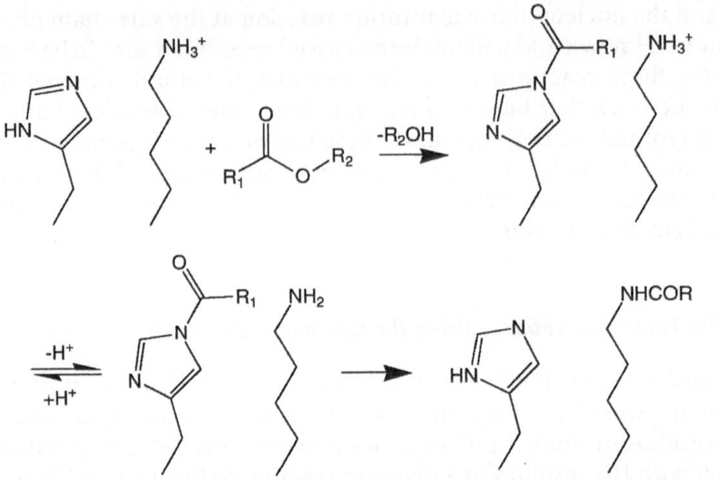

Fig. 10. The mechanism of site-selective amidation of lysine residues in folded helical structures with His (*i*) Lys (*i* + *4* or *i* − *3*) sites. The ester is typically *p*-nitrophenyl but can also be e.g. *N*-hydroxysuccinimidyl. Ornithine and diaminobutyric acid residues will also be amidated in the described positions

4.3.2
Chemoselective Ligation – Using Non-natural Amino Acids

Chemoselectivity can be used to incorporate new functionality when due to the properties of its functional group the reactivity of an amino acid side chain is several orders of magnitude larger than that of other residues. In contrast to the use of cooperative sites, described in the preceding section, this approach to the functionalization of folded proteins requires the use of non-natural amino acids. The aminooxy function [28, 29, 41] has been successfully incorporated into amino acids and used to form oximes with aldehyde groups of carbohydrates and lipids. Hydroxylamine derivatives form oximes readily in irreversible reactions and orthogonal protection group strategies are available that make it possible to regiospecifically introduce several functions sequentially in a polypeptide or protein [41]. Oxime formation appears to be insensitive to primary and secondary structure and the approach is therefore highly versatile (Fig. 11).

Chemoselective ligation of peptides using the free amino terminal as nucleophile and a thiobenzyl thioester *C*-terminal amino acid as electrophile provides an efficient approach to the synthesis of proteins with several hundred residues [30]. From this perspective the introduction of non-natural amino acids into proteins becomes a possibility rather than a problem and chemoselective ligation is thus a prospect for the future for the incorporation of new functionality.

The introduction of carbohydrates using chemoselective ligation led to the formation of a new class of glycoconjugates [66] in good yields in one-pot reac-

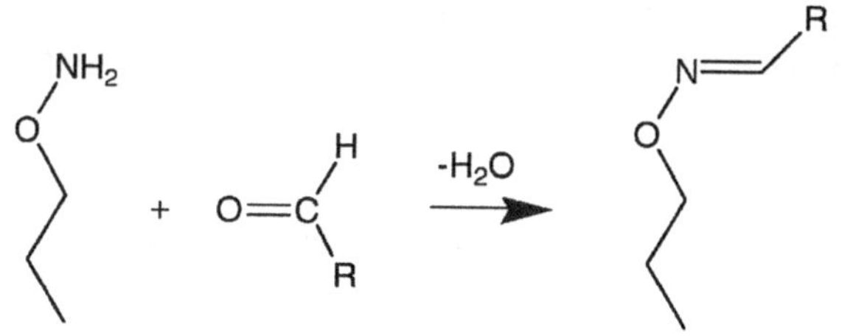

Fig. 11. Mechanism of chemoselective ligation using aminooxy functionalized residues

tions and the use of this technique for the introduction of new functionality using more accessible aldehydes therefore shows great promise.

5
Functionalized Designed Folded Proteins

The designed functionalized polypeptides and proteins that have been reported so far are mainly new catalysts [11–13], metalloproteins [14–18], heme-binding proteins [19–23] and, in a few cases, glycoproteins [65, 66]. They range in size from the 14-residue catalyst oxaldie 1 [11] to a heme-binding 126-residue protein [19]. The introduction of functional sites presents an increased level of complexity and some of the most successful examples of catalyst design are the ones where the function of the reactive sites does not depend on the exact location of several residues simultaneously. While this may discourage some, impressive results have been obtained by designed proteins that do not have native-like structures and that function without perfect structural organization. The demonstration of novel functions is in itself an important goal and the optimization of selectivity and efficiency will surely follow.

It may appear that the flexibility of the motifs is the most limiting factor but, although there is a price to be paid in terms of activation entropy when flexible side chains must come together to form a well-defined transition state, a more serious problem is probably the crudeness of the templates that can now be designed. Ideally each reactive site residue should be located in a separate secondary structure so that the geometry can be optimized in very small steps, but this is clearly not possible in simple motifs. On the other hand, as shown below, it is not necessary to rely on the positions of individual side chains when the hydrophobic effect is exploited for binding or when only a single residue forms a covalent bond to the substrate. There is therefore room for some optimism in expecting that there will be selected examples of designed catalysts that show very impressive performances.

5.1
Designed Catalysts

The problem of catalysis is very challenging in protein design and also very tempting as it tests the understanding of the folding problem as well as the understanding of enzyme catalysis. Apart from the incorporation of cofactors, a subject that is not covered here, there have been essentially three successful catalyst designs reported so far that give rise to rate enhancements of three orders of magnitude or more, and substrate discrimination, and where the relationship between structure and function is understood, at least to some extent [11–13]. A catalytically active designed four-helix bundle has also been reported that hydrolyzed p-nitrophenyl esters with substantial rate enhancements [67]. The design was based on the active site of the serine proteases and included the introduction of a catalytic triad and an oxyanion hole. The results were later partially retracted [68] although it did show reactivity towards p-nitrophenyl acetate perhaps due to the large number of lysine residues in the sequence.

This catalyst is nevertheless interesting as it attempts to mimic one of the most famous catalytic entities in enzymology, the Asp, His, Ser triad. Many attempts have been made to repeat the activity of the triad in model peptides but with little success and some of the possible reasons are of general interest in catalyst design. In the active site of chymotrypsin, serine is the nucleophile that forms the tetrahedral intermediate with the ester or the amide substrate under the release of the leaving group, and it is assisted by the general-base catalyst histidine. In aqueous solution around neutral pH His is a much better nucleophile than Ser so the prospect of mimicking the basic function of the triad is, at best slim. It is much more likely that His will be the nucleophile that reacts with the substrate to form the acyl intermediate. Furthermore, the organization of the three residues, Asp, His and Ser, in their correct geometries simultaneously has only a low probability of success because in the absence of conformational constraints there is a very large number of non-reactive conformations that compete. Finally, the organization of a triad requires that a template molecule is available where all three residues can be positioned in a way that is compatible with optimum distances for inter-residue proton-transfer reactions and this is difficult in simple constructs.

The design of a functional protease mimic thus has to be considered to be too difficult a problem for the time being. In contrast, the successful design strategies to date are based on relatively simple concepts that do not require several residues to be in the correct conformations simultaneously and where the chemistries and template structures are compatible with the intended functions.

5.1.1
The Catalysis of Decarboxylation

The first designed catalyst where there was some understanding of the relationship between structure and function was oxaldie 1, a 14-residue peptide that folds in solution to form helical bundles [11] (Fig. 12). Oxaldie 1 was designed to catalyze the decarboxylation of oxaloacetate, the α-keto acid of aspartic acid, via a mechanism where a primary amine reacts with the ketone carbonyl group to form a carbinolamine that is decarboxylated to form pyruvate. The reaction is pK_a dependent and proceeds faster the lower the pK_a of the primary amine if the reaction is carried out at a pH that is lower than the pK_a of the reactive amine. The sequence contains five lysine residues that in the folded state form

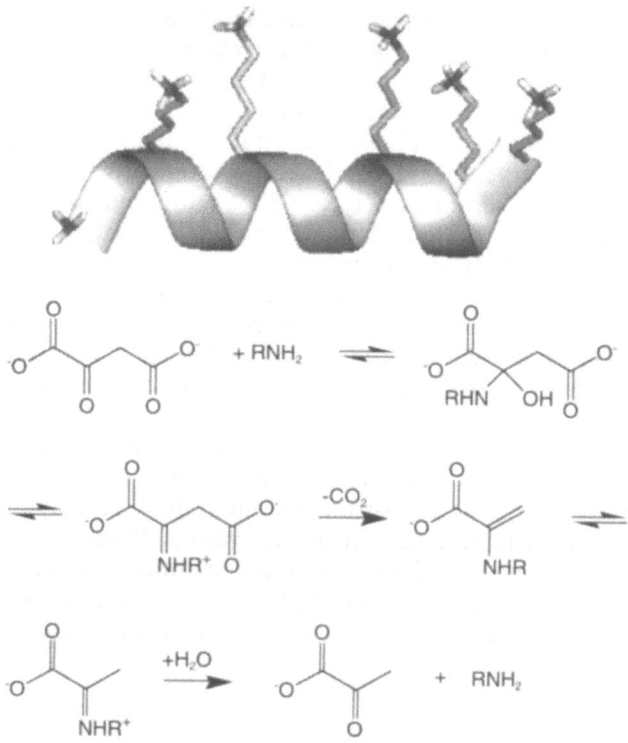

Fig. 12. Reaction mechanism of oxaldie 1 catalyzed decarboxylation of oxaloacetate. The organization of several flanking lysine residues depresses the pK_a of the catalytically active Lys side chain. The free amino terminal of the folded peptide is the most reactive residue because its pK_a is depressed by the positive end of the helix dipole

one face of the helix and the "middle" lysine is therefore surrounded on both sides by positively charged residues with the design idea that they should depress the pK_a and increase the reactivity of the "middle" lysine.

The overall reactivity of oxaldie 1 towards oxaloacetate in the formation of the reaction product is a factor of 400 times larger than that of butylamine. It follows saturation kinetics and k_{cat}/K_M is 0.47 s^{-1} M^{-1}. The function of the catalyst depends on the pK_a of the reactive primary amine and oxaldie 1 has a free N-terminal amine with a pK_a of 7.2. In order to probe the role of the amino terminal it was blocked by acetylation to form oxaldie 2 and the measured pK_a of oxaldie 2 was 8.9, which shows that the pK_a of the reactive lysine was depressed by almost two pK_a units. The reactivity of oxaldie 2 was lower than that of oxaldie 1, k_{cat}/K_M is 0.15 s^{-1} M^{-1} a factor of three less than that of oxaldie 1, and it can thus be concluded that two thirds of the reactivity of oxaldie 1 follows the amino terminal pathway, but that oxaldie 2 is nevertheless a respectable catalyst with rate enhancements in the order of two orders of magnitude over that of the butylamine-catalyzed reaction.

The observed reactivity difference where a factor of two in reactivity arises from a pK_a difference of at least 1.7 pK_a units is relatively modest and the reason for the catalytic efficiency must also be due to efficient transition state binding of the substrate by the catalyst. This is not surprising in view of the fact that this corresponds to the binding of a dianion substrate by penta- or hexacation catalysts. The modest observed pK_a dependence, however, also suggests that more than one pathway may be available in oxaldie 1 and 2 due to the presence of several lysine side chains. The elucidation of the detailed reaction mechanism is a problem worth pursuing because the catalyzed reaction shows great promise in providing an understanding of how multistep reactions are catalyzed and where enzymes derive their catalytic energy from. In particular, there is reason to expect that a thorough understanding can be obtained of how the catalysis of various reaction steps can be balanced so that the overall free energy of activation is optimized.

The most interesting design idea is perhaps that of the choice of the reaction under study. In the design of catalysts it is tempting to maximize the binding energies between catalyst and substrate. Product inhibition may therefore become a hurdle in the engineering of turnover systems. In the oxaldie-catalyzed decarboxylation reaction the substrate is a dianion under the experimental conditions whereas the product is a monoanion and there is little risk of product inhibition because the substrate will be bound substantially stronger than the product by the catalyst.

5.1.2
The Catalysis of Peptide Ligation

A catalyst designed according to very different principles was shown to enhance the reaction rate of a peptide ligation reaction by more than three orders of magnitude by exploiting the hydrophobic binding energy to organize two reactants on the surface of a folded helix [12] (Fig. 13). The catalyst was a 33-residue peptide with some helix propensity that showed only modest degrees of helix formation in aqueous solution. The catalyzed reaction is the formation of a peptide from peptide fragments where the C-terminal of the "electrophile" peptide is a thioester and the N-terminal amino acid of the "nucleophile" peptide is a cysteine. The reaction proceeds via the attack of the cysteine side chain on the C-terminal thioester to form a new thioester which rearranges to form the amide bond [30]. If the reactant peptides are amphiphilic they can assemble on the 33-residue amphiphilic peptide and react with enhanced rates due to increased proximity. This was demonstrated to be the case and the rate of the peptide-catalyzed reaction was found to be 4100 times larger than that of the uncatalyzed reaction.

The binding of the reactant peptides is mainly controlled by hydrophobic interactions between catalyst and reactants and it is likely for entropy reasons that the product peptide will bind stronger than each of the reactant peptides. It is with a remarkable degree of success that the helical propensity of the 33-residue peptide is balanced in a way that minimizes product inhibition and catalyst aggregation while the fraction of folded peptide is large enough to generate efficient catalysis. Selectivity is ensured by introducing charged residues in posi-

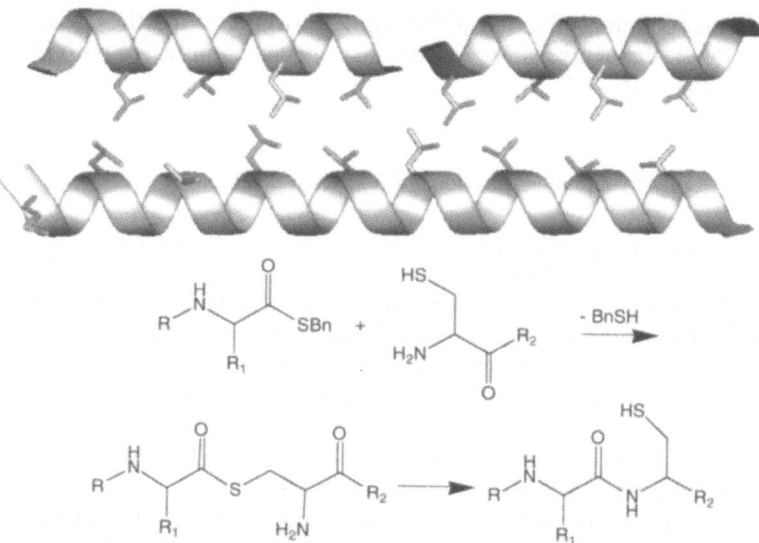

Fig. 13. Reaction mechanism of 33-residue amphiphilic helix catalyzed chemoselective ligation of helical peptides. The function of the catalyst is to organize the amphiphilic peptide reactants by hydrophobic forces on the surface of the folded helix

tions that flank the hydrophobic patch of the 33-residue peptide and the catalyst provides an interesting example of a reaction where the requirements for the exact location of the residues that perform the function are not prohibitively demanding. There are several hydrophobic residues that form the binding site and the binding of the substrates does not therefore depend on the specific interaction with a single residue.

The most important result from this investigation is perhaps the demonstration that hydrophobic interactions between amphiphilic helices and substrates in aqueous solution are strong enough to ensure sufficient complexation and efficient catalysis. The exploitation of this important discovery can perhaps be used to catalyze many other types of reactions between hydrophobic or partially hydrophobic reactants by increasing the proximity on the surface of folded polypeptides. The rates may become even more impressive when residues capable of general-acid, general-base or nucleophilic catalysis are incorporated in the catalyst even if this will clearly lead to higher levels of complexity. Possible problems of product inhibition may also be less pronounced in reactions where hydrophobic reactants are transformed into less hydrophobic products, for example, by the use of hydrophobic leaving groups in substitution reactions.

5.1.3
Self-Replicating Peptides

The machinery of the amphiphilic polypeptide templates has been successfully used to engineer self-replicating peptides. The principles described above are sufficient to control replication if the peptide fragments are the constituents of the peptide catalyst. Impressive results have been accomplished with pH control and high selectivity [69 – 72].

5.1.4
The Catalysis of Acyl-Transfer Reactions of Reactive Esters

Designed histidine-based four-helix bundle proteins have been shown to catalyze the reactions of p-nitrophenyl esters [13]. The reactivity of histidine is due to its imidazoyl side chain that reacts with active esters in a two-step reaction. In the first and rate-limiting step the imidazoyl residue reacts with the ester to form an acyl intermediate under the release of p-nitrophenol and in the second step the acyl intermediate reacts with the most potent nucleophile to form the reaction products.

The 42-residue peptide KO-42 folds in solution into a hairpin helix-loop-helix motif that dimerizes to form a four-helix bundle. On the surface of the folded motif there are six histidines with assigned pK_a values in the range 5.2 to 7.2 (Fig. 1) and the second-order rate constant for the hydrolysis of mono-p-nitrophenyl fumarate is 1140 times larger than that of the 4-methylimidazole-catalyzed reaction at pH 4.1 and 290 K [13]. The reaction mechanism was found to be pH dependent as the kinetic solvent isotope effect was 2.0 at pH 4.7 and 1.0 at pH 6.1 and the pH dependence showed that the reaction rate depended on residues in their unprotonated form with pK_a values around 5. It was thus established that there are functional cooperative reactive sites that contain protonated and unprotonated His residues.

In subsequent investigations of the function of KO-42 it was demonstrated that two-residue sites in helical segments are the basic catalytic entities and that their reactivity is due to cooperative nucleophilic and general-acid catalysis that is strongly dependent on the pK_a of the participating histidines [61]. In the process, rules were discovered for the control of histidine pK_a and reactivity by the introduction of flanking charged residues [60]. The sites are capable of cooperative catalysis and can be supplemented with flanking arginines, lysines and histidines to engineer catalysts capable of substrate discrimination and chiral recognition. Recognition of anionic and hydrophobic substituents was demonstrated in systematic studies using a number of modified peptides [43] (Fig. 14).

The helix-loop-helix dimer MNKR had a four-residue site consisting of two His, one Lys and one Arg residues and showed saturation kinetics with a K_M of 1 mM whereas KO-42 showed no sign of saturation kinetics. The rational introduction of residues that provided efficient transition state binding evidently led to binding of the ground state also [43]! Here, too, it was demonstrated that the binding energies available on the surface of folded polypeptides are large enough to provide efficient complexation for catalysis whereas the optimal organization for the side chains of the residues that are involved in bond making and breaking is hard to achieve.

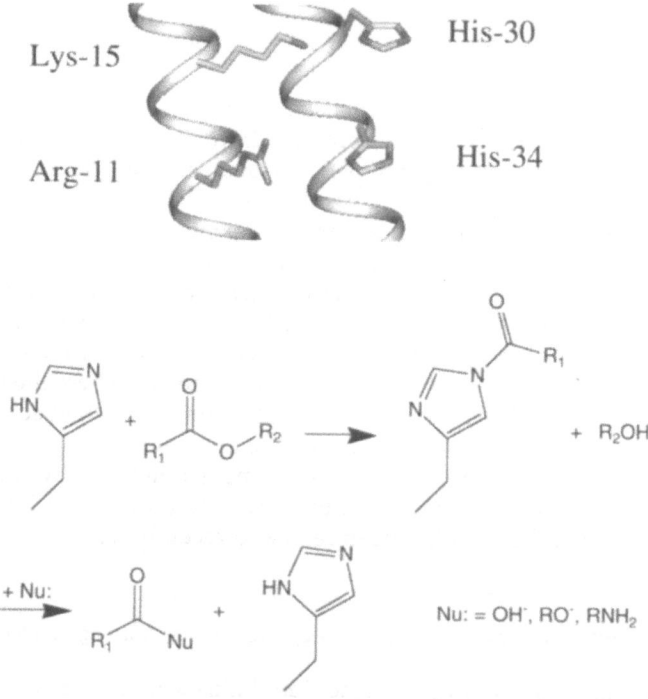

Fig. 14. Reaction mechanism of histidine-catalyzed acyl transfer of reactive esters. The excised reactive site is part of a four-helix bundle motif and is capable of substrate recognition and rate enhancements of approximately three orders of magnitude

5.1.5
Binding versus Reactivity in Designed Catalysts

The lessons that have been learnt so far from the reactivity of these designed polypeptide catalysts are that it is now possible to introduce by rational design all the complexities of native enzymes and that the binding energies that can be obtained are comparable. Nucleophilic and general-acid catalysis, substrate discrimination, saturation kinetics and the selective recognition of hydrophobic and anionic substituents have been demonstrated. Difficulties remain in optimizing geometries and perhaps in limiting the conformational freedom of the side chains of the reactive site. The problems lie in optimizing k_{cat} rather than K_M and from this perspective the design of protein-like cavities is not necessary. However, the crudeness of the relatively simple constructs that have so far been available as templates probably limits the reactive site geometries that can be engineered. Future developments will most likely have to deal with obtaining templates where there is more design flexibility.

5.2
Designed Metalloproteins

A number of laboratories have focused on the design of metallo-binding proteins because functional metalloproteins have many important functions and metal coordination can in principle control structure. The difficulties involved in organizing the metal-binding residues as well as incorporating shape complementarity in the hydrophobic core are formidable and have been discussed above. The design of metalloproteins is therefore a severe test of the understanding of protein folding. The objectives of introducing metal ions ranges from the mimicking of naturally occurring metal-binding sites to exploiting the metal to ligand coordination to control folding. Recently, however, a major advance in the design of metallo-binding peptides has been reported where the binding energy of the folded peptide was used to control the metal to ligand coordination [14]. This is an important first step towards the engineering of truly novel metalloproteins. Designed native-like metal-binding geometries have mainly been the target in attempting to redesign natural proteins, an area that is beyond the scope of this article, but that has been reviewed recently [73].

5.2.1
Control of Metal to Ligand Coordination by the Peptide Binding Energy

The binding of mercury by cysteine side chains would be expected to take place with a coordination number of 2, which is the preferred geometry, although tricoordinated complexes have been reported. Introduction of a Cys residue instead of a Leu (L16 C) in one of the hydrophobic positions in one of the heptad repeats of a 30-residue helical peptide led to a peptide that bound Hg in the hydrophobic interior of the three-helix coiled coil [14]. The coordination of the mercury ion was determined by the nature of the peptide assembly, at peptide concentrations where the coiled coil was predominantly dimeric, dicoordinated mercury was observed, whereas at peptide concentrations where the three-helix coiled coil dominated an unusual three coordination of mercury was identified by ^{199}Hg NMR spectroscopy and by EXAFS. An important feature of the design is that each coordinating residue is located in a separate secondary structure and on the "inside" of the folded structure. This expands the range of geometries accessible for functionalization considerably. The demonstration that metal ions can be introduced in peptide environments that are protected from solvent also suggests that it is now possible to make water-sensitive coordination complexes and oxidation states "water soluble". The observation of peptide-controlled metal coordination provides some hope that designed functional metalloproteins with novel properties may be available in the near future.

5.2.2
Controlling Peptide Folding by Metal to Ligand Coordination

The alternate approach is to make use of the strong metal to ligand bonds to control the folding of peptides and proteins [15–18, 74]. In this way helical peptides have been shown to fold in the presence of metals [16, 74], in spite of intrinsically low helical propensity. Furthermore, Zn-binding sites introduced into a designed four-helix bundle protein were shown to control the topology of the folded state by excluding large populations of poorly defined folds [17]. This is a rare example where a designed protein has been shown to approach a more native-like state upon metal binding, perhaps because the initial state had the characteristics of a molten globule. In the cases where the apoprotein has a more defined tertiary structure the opposite is often found. The initiation of folding and the directionality of the secondary structure elements were successfully controlled by the metal in a designed three-helix bundle protein and in a designed β-sheet demonstrating the usefulness of metals in peptide assembly [16, 18].

5.3
Designed Heme-Binding Proteins

Heme-binding proteins are also metal-binding proteins but the explosion of successful designs and strategies in this area warrants the presentation of this field in a separate section. Hemes perform an array of functions in nature such as oxygen transport, electron transport, oxidation and reduction reactions and in addition to shedding light on the function of the naturally occurring heme-proteins a number of applications can be envisioned such as the engineering of new catalysts and biosensors. The design problem is formidable because not only does the protein bind the metal but the largely hydrophobic porphyrin superstructure also has to be accommodated. Ferric hemes are bound by His residues and the basic design therefore includes two histidines per heme, but in spite of the fact that iron binding by imidazole is strong and cooperative the incorporation of ferrous hemes requires more than the availability of histidine residues.

In an impressive achievement, four iron porphyrins were incorporated into one designed protein and the measured redox potentials of each one varied significantly, although not as much as those of native heme proteins [19]. This highly sophisticated model system shows great promise with regards to providing an understanding of how the microenvironment of the proteins can be used to control the electrochemical properties of the prosthetic group. The apoprotein has the characteristics of non-native proteins but subsequent redesigns show well-dispersed NMR spectra [7]. It will be interesting to see whether it will be possible to retain the well-defined tertiary structure upon incorporation of the heme groups and whether the effects on redox potentials will be more significant as a consequence of a more native-like fold. Then it will be possible to resolve the intriguing question of whether the interior of a molten globule-like protein is transiently accessible to solvent water and whether more efficient expulsion of water molecules is the key to even better control of the redox properties of model heme proteins.

In a further demonstration of the scope of the four-helix bundle *maquette* further complexity was added by the addition of both the flavin cofactor and heme groups [75]. Photoreduction of the hemes was successfully demon-

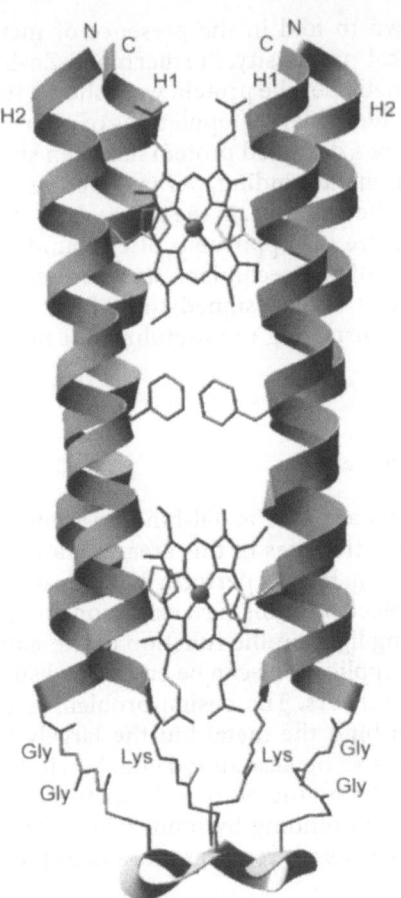

Fig. 15. Modelled structure of MOP1 bis-heme complex. Reproduced with permission from J Am Chem Soc (1998) 120:468. (© 1998 ACS)

strated by electron transfer between light-activated flavins and the iron porphyrins.

In order to circumvent the sometimes cumbersome development of folded designed proteins with defined tertiary structures the TASP approach was applied to the engineering of a four-helix bundle heme-binding protein [20] (Fig. 15). The design of the helical segments were based on heme-binding helical segments from the literature [19] and on the sequence of a naturally occurring protein. A cyclic decapeptide with four individually protected Cys residues served as the template and the helices were linked to the Cys side chains by reaction with incorporated bromoacetyl groups. At the *N*-terminal the bromoacetyl group was linked via an amide bond to the amino group, whereas at the *C*-terminal it was linked to the side chain of a lysine residue, a strategy that made it possible to control the directionality of the helical segments. The model protein showed two different midpoint potentials and when NMR spectroscopic investigations are available of the apoprotein as well as of the functional TASP protein an interesting evaluation will be possible of the TASP versus the linear approach to protein functionalization. In particular, the issue of transient solvent accessibility to the protein interior may be an important one. In a further development of the heme-binding TASP molecules light-induced electron transfer was accomplished in a single heme four-helix bundle protein where a ruthenium complex had been covalently linked by chemoselective ligation [55].

The strong binding of sulfur to gold provides an efficient way of functionalizing surfaces by peptides and proteins. The TASP strategy is highly suitable for this purpose since the immobilization can be controlled so that a homogeneous surface is obtained. This is in contrast to immobilization using only the hydrophobic interactions between protein and the surface or the non-selective chemical methods that are commonly used to bind proteins to surfaces. In this way designed TASP heme proteins were bound to a gold electrode and electrical contact between the electrode and the protein was established [76].

While these complex model heme proteins have a large potential for functionalization, an interesting approach that is very different has been taken by other workers in that the heme itself functions as the template in the formation of folded peptides. In these models peptide–peptide interactions are minimized and the driving force for folding appears to be the interactions between porphyrin and the hydrophobic faces of the amphiphilic peptides. The amino acid sequences are too small to permit peptide–peptide contacts as they are separated by the tetrapyrrole residue. These peptide heme conjugates show well-resolved NMR spectra and thus well-defined folds and the relationship between structure and function can probably be determined in great detail when functions have been demonstrated [22, 23, 77]. They are therefore important model systems that complement the more complex proteins described above.

5.4
Designed Glycoproteins

Glycoproteins play an important role in, for example, immunology and the elucidation of the functions of the carbohydrate residues are therefore the focus of great current efforts. The protein-bound carbohydrates perform an array of functions ranging from protection against proteolysis to the complex interplay with receptors and the discrimination of these functions in simple model systems can perhaps aid in a more detailed understanding of protein–carbohydrate interactions.

The introduction of sugar amino acids through peptide synthesis is often cumbersome due to the complex protection group strategies that are necessary in the synthesis of the amino acids while the post-synthetic incorporation of carbohydrates can be accomplished with comparative ease. Chemoselective ligation has been used to incorporate a number of carbohydrates into a somatostatin analogue that carries artificial amino acids with a side chain that has an aminooxy function [66]. In combination with orthogonal protection group strategies this approach allows the selective introduction of several carbohydrates in folded proteins or polypeptides and although few attempts have been reported the mimicking of natural antigens or receptors may now be possible in designed glycoproteins.

In an alternative approach the site-selective functionalization reaction has been used to incorporate a galactose derivative into a folded four-helix bundle protein and the effect of glycosylation on the structure of the folded protein has been identified [65]. The unfunctionalized designed four-helix bundle did not have a well-defined tertiary structure but the introduction of the sugar improved the helical content and reduced the rate of conformational exchange. Glycosylation may therefore play a role in the maturation of poorly folded proteins.

6
Conclusions

The *de novo* design of proteins using non-natural sequences and an understanding of the principles that control protein folding in nature has now reached a level where the introduction of function is a viable alternative to the re-engineering of naturally occurring proteins. The driving force for this development is the expectation that new functions require new shapes and that the elucidation of the principles that control the function of the biomolecules is difficult since the variety of tasks that they perform may mask individual functions. The engineering of a functional protein is most often based on the assumption that a template protein is available where a few modifications of the amino acid functions will be sufficient to create reactive or binding sites for functional groups. The rapid expansion of the number of designed structural motifs in recent years is therefore gratifying because it provides a large number of different geometries that can be explored in the pursuit of function. Several questions remain with regards to protein tolerance towards structural modification: with regards to conformational stability and function and as to whether the templates available so far can be fine tuned to, for example, enzyme-like perfection. Nevertheless, the range of functions that have been demonstrated and reviewed here represents an impressive increase in new shapes and new functions in a very short time and the prospect for the future is therefore one of great promise.

Acknowledgements. I am indebted to Dr Gabriele Tuchscherer for accurate references on TASP molecules and chemoselective ligation and to Dr Murray Goodman for stimulating discussions on designed folded proteins.

7
References

1. Hill RB, DeGrado WF (1998) J Am Chem Soc 120:1138
2. Brive L, Dolphin GT, Baltzer L (1997) J Am Chem Soc 119:8598
3. Schafmeister CE, LaPorte SL, Miercke LJW, Stroud RM (1997) Nature Struct Biol 4:1039
4. Roy S, Ratnaswami G, Boice JA, Fairman R, McLendon G, Hecht MH (1997) J Am Chem Soc 119:5302

5. Struthers MD, Cheng RP, Imperiali B (1996) Science 271:342
6. Betz SF, DeGrado WF (1996) Biochemistry 35:6955
7. Gibney BR, Rabanal F, Skalicky JJ, Wand AJ, Dutton PL (1997) J Am Chem Soc 119:2323
8. Jefferson EA, Locardi E, Goodman M (1998) J Am Chem Soc 120:7420
9. Kortemme T, Ramirez-Alvarado M, Serrano L (1998) Science 281:253
10. Schenk HL, Gellman SH (1998) J Am Chem Soc 120:4869
11. Johnsson K, Allemann RK, Widmer H, Benner SA (1993) Nature 365:530
12. Severin K, Lee DH, Kennan AJ, Ghadiri MR (1997) Nature 389:706
13. Broo KS, Brive L, Ahlberg P, Baltzer L (1997) J Am Chem Soc 119:11362
14. Dieckmann GR, McRorie DK, Tierney DL, Utschig LM, Singer CP, O'Halloran TV, Penner-Hahn JE, DeGrado WF, Pecoraro VL (1997) J Am Chem Soc 119:6195
15. Kohn WD, Kay CM, Sykes BD, Hodges RS (1998) J Am Chem Soc 120:1124
16. Ghadiri MR, Case MA (1993) Angew Chem Int Ed 32:1594
17. Handel TM, Williams SA, DeGrado WF (1993) Science 261:879
18. Schneider JP, Kelly JW (1995) J Am Chem Soc 117:2533
19. Gibney BR, Rabanal F, Reddy KS, Dutton PL (1998) Biochemistry 37:4635
20. Rau HK, Haehnel W (1998) J Am Chem Soc 120:468
21. Rojas NRL, Kamtekar S, Simons CT, McLean JE, Vogel KM, Spiro TG, Farid RS, Hecht MH (1997) Protein Sci 6:2512
22. Arnold PA, Shelton WR, Benson DR (1997) J Am Chem Soc 119:3181
23. Nastri F, Lombardi A, Morelli G, Maglio O, D'Auria G, Pedone C, Pavone V (1997) Chem Eur J 3:340
24. Baltzer L, Lundh A-C, Broo K, Olofsson S, Ahlberg P (1996) J Chem Soc Perkin Trans 2 1671
25. Broo K, Brive L, Lundh A-C, Ahlberg P, Baltzer L (1996) J Am Chem Soc 118:8172
26. Kemp O, Carey R (1991) Tetrahedron Lett 32:2845
27. Dawson PE, Kent SBH (1993) J Am Chem Soc 115:7263
28. Tuchscherer G (1993) Tetrahedron Lett 34:8419
29. Vilaseca LA, Rose K, Werlen R, Meunir A, Offord RE, Nichols CL, Scott WL (1993) Bioconjugate Chem 4:515
30. Dawson PE, Muir TW, Clark-Lewis I, Kent SBH (1994) Science 266:776
31. Mutter M (1988) In: Marshall G (ed) Peptides, chemistry and biology. p 349
32. Mutter M, Tuchscherer G (1988) Makromol Chem Rapid Commun 9:437
33. Mutter M, Vuillemier S (1989) Angew Chem Int Ed 28:535
34. Mutter M, Dumy P, Garrouste P, Lehmann C, Mathieu M, Peggion C, Peluso S, Razaname A, Tuchscherer G (1996) Angew Chem Int Ed 35:1482
35. Mutter M, Tuchscherer G (1997) Cell Mol Life Sci 53:851
36. Kamtekar S, Schiffer JM, Xiong H, Babik JM, Hecht MH (1993) Science 262:1680
37. Bryson JW, Betz SF, Lu HS, Suich DJ, Zhou HX, O'Neil KT, DeGrado WF (1995) Science 270:935
38. Minor DL Jr, Kim PS (1994) Nature 367:660
39. Smith CK, Regan L (1995) Science 270:980
40. Kemp DS, Petrakis KS (1981) J Org Chem 46:5140
41. Nyanguile O, Mutter M, Tuchscherer G (1994) Lett Pept Sci 1:9
42. Olofsson S, Johansson G, Baltzer L (1995) J Chem Soc Perkin Trans 2 2047
43. Broo KS, Nilsson H, Nilsson J, Baltzer L (1998) J Am Chem Soc 120:10287
44. Betz SF, Raleigh DP, DeGrado WF (1993) Curr Opin Struct Biol 3:601
45. Ho SP, DeGrado WF (1987) J Am Chem Soc 109:6751
46. Raleigh DP, DeGrado WF (1992) J Am Chem Soc 114:10079
47. Raleigh DP, Betz SF, DeGrado WF (1995) J Am Chem Soc 117:7558
48. Dolphin GT, Baltzer L (1997) Folding Des 2:319
49. Kuroda Y, Nakai T, Ohkubo T (1994) J Mol Biol 236:862
50. Fezoui Y, Connolly PJ, Osterhout JJ (1997) Protein Sci 6:1869
51. Olofsson S, Baltzer L (1996) Folding Des 1:347
52. Struthers MD, Cheng RP, Imperiali B (1996) J Am Chem Soc 118:3073
53. Struthers M, Ottesen JJ, Imperiali, B (1998) Folding Des 3:95

54. Kohn WD, Hodges, RS (1998) Trends Biotechnol 16:379
55. Rau HK, DeJonge N, Haehnel W (1998) Proc Natl Acad Sci 95:11526
56. Goodman M, Feng Y, Melacini G, Taulane JP (1996) J Am Chem Soc 118:5156
57. Goodman M, Melacini G, Feng Y (1996) J Am Chem Soc 118:10928
58. Melacini G, Feng Y, Goodman M (1996) J Am Chem Soc 118:10359
59. Melacini G, Feng Y, Goodman M (1997) Biochemistry 36:8725
60. Broo KS, Brive L, Sott RS, Baltzer L (1998) Folding Des 3:303
61. Broo KS, Nilsson H, Nilsson J, Flodberg A, Baltzer L (1998) J Am Chem Soc 120:4063
62. Baltzer L, Broo KS, Nilsson H, Nilsson J (1998) Bioorg and Med Chem in press
63. Broo K, Allert M, Andersson L, Erlandsson P, Stenhagen G, Wigström J, Ahlberg P, Baltzer L (1997) J Chem Soc Perkin Trans 2 397
64. Allert M, Kjellstrand M, Broo K, Nilsson Å, Baltzer L (1998) J Chem Soc Chem Commun 1547
65. Andersson L, Stenhagen G, Baltzer L (1998) J Org Chem 63:1366
66. Cervigni SE, Dumy P, Mutter M (1996) Angew Chem Int Ed 35:1230
67. Hahn KW, Klis WA, Stewart JM (1990) Science 248:1544
68. Corey MJ, Hallakova E, Pugh K, Stewart JM (1994) Appl Biochem Biotechnol 47:199
69. Severin K, Lee DH, Martinez JA, Ghadiri MR (1997) Chem Eur J 3:1017
70. Lee DH, Severin K, Yokobayashi Y, Ghadiri MR (1997) Nature 390:591
71. Yao S, Ghosh I, Zutshi R, Chmielewski J (1997) J Am Chem Soc 119:10559
72. Yao S, Ghosh I, Zutshi R, Chmielewski J (1998) Angew Chem Int Ed 37:478
73. Lu Y, Valentine JS (1997) Curr Opin Struct Biol 7:495
74. Ruan F, Chen Y, Hopkins PB (1990) J Am Chem Soc 112:9403
75. Sharp RE, Moser CC, Rabanal F, Dutton PL (1998) Proc Natl Acad Sci 95:10465
76. Katz E, Heleg-Shabtai V, Willner I, Rau HK, Haehnel W (1998) Angew Chem Int Ed in press
77. D'Auria G, Maglio O, Nastri F, Lombardi A, Mazzeo M, Morelli G, Paolillo L, Pedone C, Pavone V (1997) Chem Eur J 3:350
78. Powers R, Garrett DS, March CJ, Frieden EA, Gronenborn AM, Clore CM (1992) Biochemistry 31:4334
79. Abola EE, Sussman JL, Prilusky J, Manning NO (1997) Protein data bank archives of three-dimensional macromolecular structures. In: Carter CW Jr, Sweet RM (eds) Methods in enzymology, vol 277. Academic Press, San Diego, pp 556–571
80. Abola EE, Bernstein FC, Bryant SH, Koetzle TF, Weng J (1987) In: Allen FH, Bergerhoff G, Sievers R (eds) Protein data bank in crystallographic databases-information content, software systems, scientific applications. Data Commission of the International Union of Crystallography, Bonn Cambridge Chester, pp 107–132

Incorporation of Noncoded Amino Acids by In Vitro Protein Biosynthesis

Marcella A. Gilmore[1] · Lance E. Steward[2] · A. Richard Chamberlin[1,*]

[1] Department of Chemistry, University of California, Irvine, CA 92697, USA.
E-mail: archambe@uci.edu

[2] *Current address:* L.E. Steward, Department of Chemistry, University of Utah, Salt Lake City, UT 84112, USA

The method of site-specific mutagenesis with noncoded amino acids using suppression of a nonsense codon by a semi-synthetic tRNA was first introduced in 1989. Initially used to probe the tolerance of the protein biosynthetic machinery for compounds other than the 20 primary amino acids, the method has since been applied to study a widely diverse range of biological problems. The ability to introduce side chains bearing subtle structural and electronic differences, fluorescent probes, isotope labels, photolabile protecting groups, chemical handles and photoactivated cross-linkers at unique sites has facilitated studies not currently accessible by other means. Improvements and alternatives to the early methodology are considered as well as some interesting recent applications.

Keywords: Noncoded amino acids, Site-directed suppression mutagenesis, In vitro protein biosynthesis.

---

* Corresponding author.

List of Abbreviations

tRNA	transfer ribonucleic acid
ER	endoplasmic reticulum
RF	release factor
E. coli	*Escherichia coli*
M. luteus	*Micrococcus luteus*
T. thermophila	*Tetrahymena thermophila*
nAChR	nicotinic acetylcholine receptor
Arg	arginine
Asp	aspartic acid
Gln	glutamine
Asn	asparagine
Gly	glycine
Phe	phenylalanine

1
Introduction

In the past two decades protein engineering has evolved into a field that spans the physical, biological, and medical sciences and today is one of the mainstays of the multibillion-dollar biotechnology industry. Much of this evolution coincided with advances in genetic engineering and recombinant DNA technologies, as well as with the development of methods for efficient and inexpensive synthesis of oligonucleotides. Following these early advances, the first report of in vitro oligonucleotide-directed mutagenesis to study protein function was made in 1982 [1] and only a few months later the term protein engineering was coined [2]. Since then, the method of site-directed mutagenesis has become one of the most powerful tools available for protein research and serves as the cornerstone of protein engineering, allowing selective modification of proteins for basic mechanistic enzymology and commercial genetic engineering.

Site-directed mutagenesis allows one or more specific amino acids in a protein to be replaced with any of the other 20 coded amino acids that are normally present in proteins [3] and has been utilized extensively to study and modify the biophysical and biochemical characteristics of a wide variety of proteins. Despite its tremendous impact, conventional site-directed mutagenesis suffers from the strict constraint that substitutions are restricted to the 20 primary amino acids. While a tremendous amount of information has been obtained from such substitutions, there are many cases in which the incorporation of a noncoded amino acid could yield more detailed information or enable researchers to investigate or engineer proteins in novel ways.

Currently a number of techniques are used to introduce noncoded amino acids into proteins. These encompass a broad range of techniques, including total synthesis of the target protein [4], semi-synthetic methods in which a synthetically prepared fragment is combined with a native polypeptide fragment(s) [4d, 5], post-translational modification by chemical or enzymatic means [6], and both in vivo and in vitro biosynthetic techniques. Each method has its advantages and limitations. At one end of this spectrum lies total synthesis of the target protein, which offers the greatest freedom in the number and type of amino acids that can be incorporated. While technological advances in peptide chemistry have made the synthesis of 30- to 50-residue peptides relatively commonplace, synthesis of longer peptides and proteins is still relatively rare and is usually accomplished through ligation of peptide segments via chemical or enzymatic methods. The difficulties inherent in the synthesis of larger proteins strongly limits the use of this approach.

At the other end of the spectrum lie in vivo techniques for expressing proteins containing noncoded amino acids [7]. In vivo expression systems range from bacterial fermentations to the administration of analogs to animals and humans. The noncoded amino acid is incubated with the cells or is supplied to the organism, which in many cases is a bacterial strain that is unable to synthesize one or more amino acids (auxotroph). These methods are subject to two critical limitations. First, if the protein has more than one residue corresponding to the amino acid analog there is little or no control over the number of natural residues that will be substituted. Even if the natural residue appears just once in the protein, a heterogeneous mixture of the mutant protein and wild-type protein containing the native residue will often be produced. Additionally, the types of amino acids that can be incorporated are limited to those that can serve as substrates for the biosynthetic machinery, in particular the tRNA synthetases, and so must be very similar in structure to the coded amino acids.

In vitro translation in cell-free extracts offers a way to circumvent some of the difficulties experienced with in vivo incorporation of noncoded amino acids while still exploiting the biosynthetic machinery, albeit with reduced efficiency. An analog can easily be substituted for any of the primary amino acids and the relative concentrations of intrinsic and noncoded amino acids can be controlled reasonably well. If the analogs must be attached to tRNA by endogenous aminoacyl synthetases, then they are still restricted to those structurally similar to coded amino acids. One technique for incorporating noncoded amino acids that has been used for many years involves modification of a functional amino acid, usually lysine or cysteine, that is already acylated to tRNA [8]. Affinity probes, fluorescent tags and biotin labels have been introduced in this manner, though the variety is necessarily limited to derivatives of a few primary amino acids. This approach can also lead to a heterogeneous product mixture, even if the protein contains a single residue of the modified type, because the discharged tRNA can be reacylated by endogenous synthetases. The introduction of site-directed suppression mutagenesis [9] facilitated the site-specific incorporation of a wider range of noncoded amino acids into peptides and proteins. The inspiration for this method was drawn from naturally occurring suppression of nonsense codons [10] and methodology for preparation of misacylated tRNAs developed in the Hecht laboratory [11], with modifications by the Brunner lab [12].

2
The Site-Directed Suppression Mutagenesis Technique

Site-directed suppression mutagenesis is conceptually simple. Specificity is achieved by replacing the codon that is to be mutated with a termination codon. During translation this codon is recognized by a suppressor tRNA engineered to contain an anticodon corresponding to the stop codon. Use of the stop codon to incorporate the desired amino acid precludes competition with naturally occurring tRNAs and thus avoids the heterogeneous product mixtures produced by other methods. Importantly, chemical aminoacylation of the suppressor tRNA bypasses the most restrictive component in the biosynthesis of proteins, the tRNA synthetases responsible for charging tRNAs with their cognate amino acids [13], permitting the introduction of structurally and functionally diverse amino acids. In the most common approach, the gene for the target protein is placed in a plasmid under the control of an inducible promoter, and the codon for the amino acid residue to be substituted is replaced with an amber termination codon by site-directed mutagenesis.

A semi-synthetic suppressor tRNA with the complementary anticodon and carrying the amino acid analog is added, along with the plasmid DNA, to a cell-free expression system, resulting in readthrough of the termination codon and production of a full-length mutant protein (Fig. 1). *Escherichia coli* S30 lysates are most often employed, but expression in rabbit reticulasyte lysates and wheat germ extracts has also been successful. A number of variations of this general scheme have been employed, both in attempts to improve protein production and because simpler alternatives sometimes exist for specific cases.

The suppressor tRNA is arguably the most critical component of this system. While different methods of preparation have been tested [14], the simplest and most widely utilized methodology has the following general features [15]: (i) a truncated, 74-nucleotide tRNA-C_{OH} missing the 3′-terminal cytidine and adenosine moieties and containing an anticodon complementary to the termination codon is prepared by runoff transcription from linearized plasmid DNA, (ii) a hybrid DNA/RNA dinucleotide (pdCpA), which offers advantages over a pure RNA dinucleotide in terms of stability and synthetic simplicity, is synthesized and mono-aminoacylated at the 2′(3′)-hydroxyl group and (iii) preparation of the aminoacyl-tRNA-dCA is completed by enzymatically ligating the aminoacyl dinucleotide to the 74-mer tRNA with T4 RNA ligase (Fig. 2). The nature of the α-amine protecting group determines whether the amine is deprotected prior to or following the ligation step.

The suppression mutagenesis technique has proven effective in highly diverse applications, from probing the synthetic capabilities of the ribosome to studying electron transfer within proteins. The types of noncoded amino acids that can be successfully introduced are equally varied, e.g. amino acid analogs exhibiting more subtle variations than are possible with the canonical twenty, fluorescent probes, photocross-linking agents, unique chemical handles for posttranslational modification, isotopic and spin labels, and protected functional groups, among others.

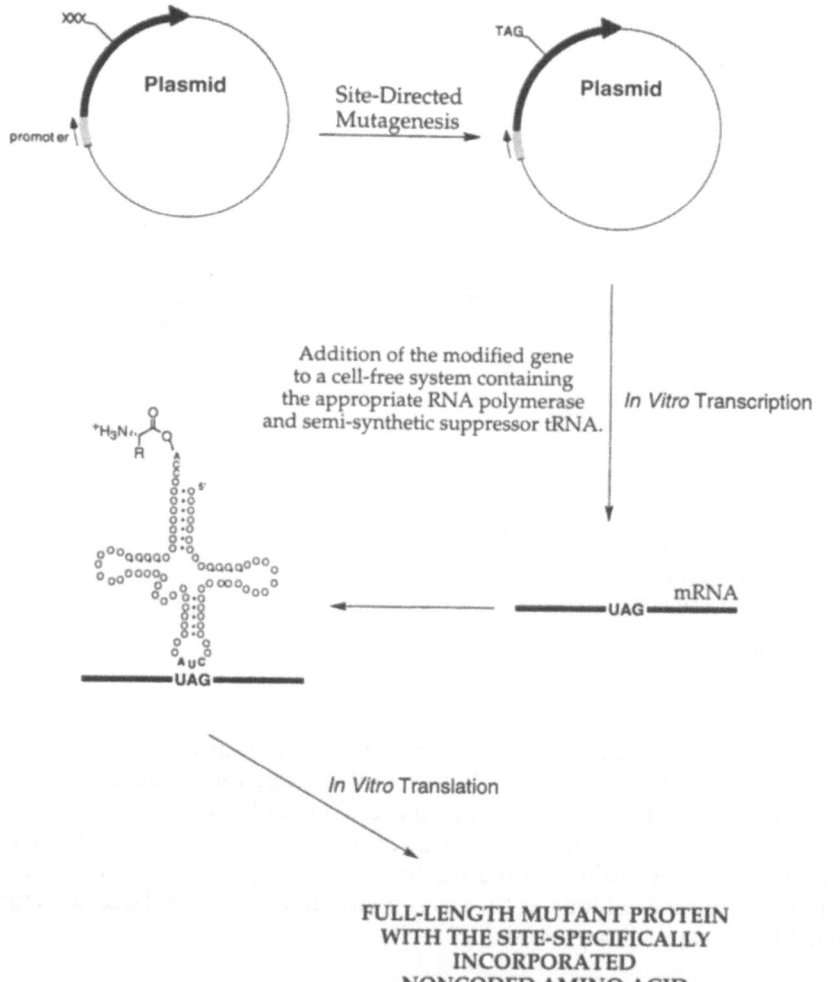

Fig. 1. The in vitro site-directed suppression mutagenesis system. Utilizing site-directed mutagenesis a specific codon within a gene under the control of an inducible promoter is converted to an amber termination codon. The gene is added to a cell-free expression system and transcription is induced in the presence of an aminoacyl suppressor tRNA, yielding protein containing the noncoded amino acid at the site corresponding to the termination codon

The principal drawbacks to the technique are its labor-intensive nature and low protein yields, which in the best cases reach only a milligram or less. Competition with release factors at the amber stop codon often results in truncated protein as the primary product [16, 17]. Suppression efficiency rates are also affected by the character of the amino acid, which determines whether it is a good substrate for the ribosome and protein elongation factors. In addition, context effects variously ascribed to the influence of neighboring mRNA

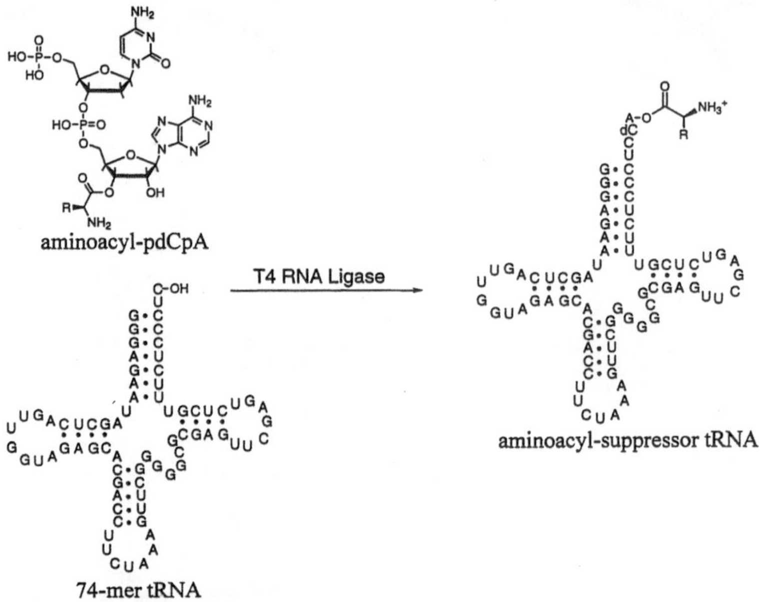

Fig. 2. Enzymatic ligation of 74-mer tRNA-C_{OH} and aminoacyl-pdCpA

sequences and to interactions of adjacent tRNAs on the surface of the ribosome [18] can dramatically affect the efficiency of expression, independent of the structure of the amino acid to be introduced. The modifications that have been made in trying to overcome these drawbacks, and some novel and exciting applications and extensions of the suppression mutagenesis technique, will be considered after a brief look at the types of amino acids and related structures amenable to protein biosynthesis.

3
The Tolerance of Ribosome-Mediated Peptide Synthesis for Noncoded Amino Acids

The earliest work with suppression mutagenesis focused on verifying the accuracy and efficiency of suppression [9, 19] and probing the extent to which structural deviations from the coded amino acids would be tolerated in ribosome-mediated peptide synthesis [20]. These studies revealed that the structures of the amino acids which can be incorporated are quite varied, as is the efficiency of their incorporation [21]. Categories of analogs that could not be incorporated at all include the D-amino acids and those with extended backbones, such as β- and γ-amino acids and dipeptides. These results were in good agreement with those obtained using other techniques [22]. Schultz and co-workers have screened a large number of noncoded amino acids and reviewed their results [15, 23.]. The selection of noncoded amino acids incorporated into proteins continues to

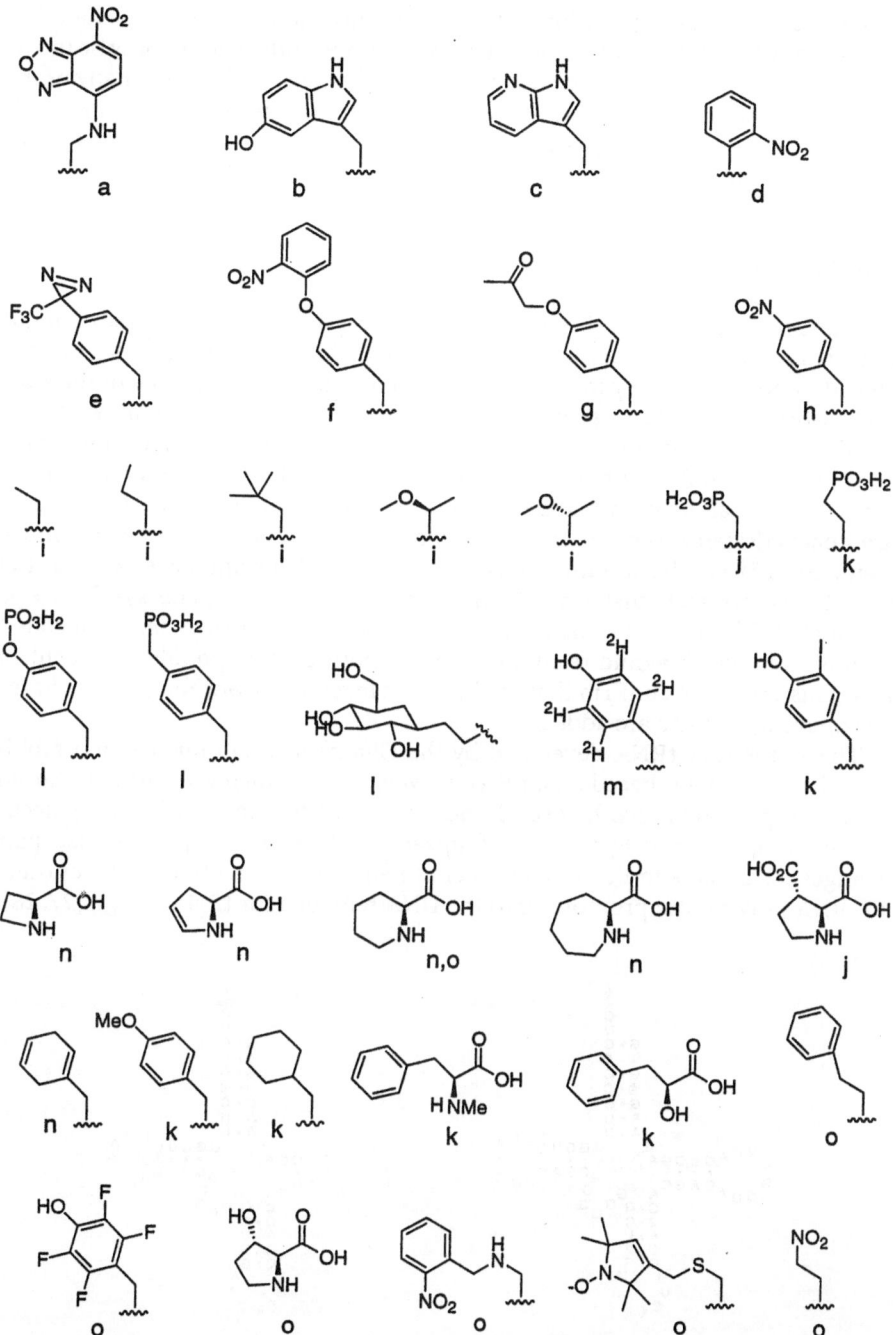

Fig. 3. A sample of amino acids and related structures that have been incorporated into peptides and proteins using site-directed suppression mutagenesis. References: (a) 63, (b) 64, (c) 23b, 29, (d) 62b, (e) 59, 60, (f) 62c, (g) 65, (h) 66, (i) 31, (j) 56b, (k) 20, (l) 56a, (m) 61c, (n) 67, (o) 23b

expand, along with applications of the technique; a sampling is included in Fig. 3. It must be noted that this is not a list of compounds that can only be incorporated with the site-directed suppression mutagenesis technique; many of the functional groups and analogs shown have been introduced by other means. It has been known for many years, for example, that ribosome-mediated ester bond formation is possible [24].

4
The Suppressor tRNA

The amber suppressor tRNAs that have been exploited for suppression mutagenesis are based upon natural tRNAs from various organisms. The primary consideration is that they must not be acylated or deacylated by any of the aminoacyl-tRNA synthetases present in the cell-free translation system. With this requirement in mind, the first suppressor tRNAs used were derived from organisms unrelated to the source of the lysate. This is not always necessary, however, since a tRNA can be modified so that it is no longer a substrate for the cognate aminoacyl tRNA synthetase. Specifically, the structural determinants or identity elements [25] can be modified (e.g. by "swapping" specific base pairs in the acceptor stem) such that interaction with the normal cognate synthetase is eliminated. To assure that the suppressor tRNAs are not reacylated with coded amino acids, which would result in heterogeneous protein populations, control reactions are run to confirm that readthrough levels are low when only uncharged suppressor tRNAs are added.

The suppressor tRNA developed by the Chamberlin lab for use in a rabbit reticulocyte lysate is based on an *E. coli* glycyl tRNA, which was initially chosen because glycyl-tRNA synthetases do not rely on a "double-sieve" editing mechanism for enzymatic hydrolysis of misacylated tRNAs [26]. Two base pair changes were made to the acceptor stem to allow incorporation of the optimal T7 RNA polymerase promoter into the DNA template for tRNA^{Gly}-C$_{OH}$ [27, 28],

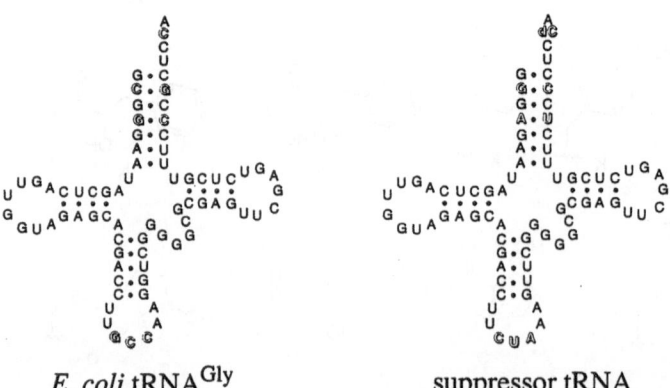

$$E.\ coli\ tRNA^{Gly} \qquad\qquad suppressor\ tRNA$$

Fig. 4. Comparison of *E. coli* tRNA^{Gly3} and semi-synthetic suppressor tRNA^{Gly}. The residues that have been changed are highlighted

and it was later found that these changes also eliminated recognition by the *E. coli* glycyl-tRNA synthetase [29]. This suppressor tRNA (Fig. 4) is now used successfully for suppression mutagenesis in *E. coli* lysates. The Schultz lab initially used a yeast tRNA^{Phe}-derived suppressor tRNA in *E. coli* lysates because it was known that yeast tRNA^{Phe} is poorly recognized by the *E. coli* phenylalanyl-tRNA synthetase and that yeast tRNA^{Phe} suppressors function efficiently in vivo [9 d and references therein]. These suppressor tRNAs are still in use, but others have been more recently employed, and two of these are discussed in the following sections.

4.1
tRNA^{Gln,CUA} from *Tetrahymena thermophila*

The Lester and Dougherty labs, which have collaborated to extend the suppression mutagenesis technique to *Xenopus* oocytes with remarkable success [30, 31], began with a suppressor tRNA ("MN3") designed for in vivo use and demonstrated that it functioned more effectively in the oocyte system than a yeast tRNA^{Phe}-derived suppressor tRNA. They have since developed an alternative suppressor based on tRNA^{Gln,CUA} from *Tetrahymena thermophila* that has proven to be considerably more versatile, efficient and accurate in the oocyte system [32], as well as showing good suppression efficiency in *E. coli* transcription-translation reactions [33].

The tRNA species from which the new suppressor tRNA is derived deviates from the universal genetic code in that it naturally introduces a glutamine in response to the UAG codon in *T. thermophila* and was known to be an efficient in vitro suppressor. Changes were made to the acceptor stem sequence which, by analogy with the interactions of *E. coli* glutaminyl-tRNA synthetase with its substrate, were expected to preclude recognition by endogenous *Xenopus* glutaminyl-tRNA synthetase. This tRNA ("THG73") and a closely related variant differing only by having an adenosine at position 73 ("THA73") were compared to MN3.

The efficiency of THG, THA and MN3 tRNAs in the oocyte system were compared by introduction of tyrosine into two different sites in the α-subunit of the nicotinic acetylcholine receptor (nAChR) [32]. Charged suppressor tRNA and mRNA were coinjected into oocytes and the suppressor efficiencies were evaluated by electrophysiological methods rather than by comparing protein yields. Efficiency was reported as the ratio of acetylcholine-induced current measured following injection of the acylated suppressor tRNA to the current measured following injection of uncharged suppressor tRNA (designated "Efficiency Ratio", Table 1). Higher efficiency ratios equate to higher confidence that measured currents are a result of receptors expressed with the correct amino acid. No current was detectable in control reactions to which no suppressor tRNA was added, indicating that readthrough by endogenous tRNAs was negligible. The current measured when uncharged suppressor tRNAs were added was therefore an indication of the extent to which the various suppressor tRNAs were acylated by endogenous tRNA synthetases, leading to synthesis of functional nAChR.

Table 1. Comparison of suppressor tRNAs in the *Xenopus* oocyte system. See text for details

UAG Position in nAChR	Suppressor tRNA	Efficiency ratio
α198	MN3	980
	THA73	280
	THG73	680
α180	MN3	1.6
	THA73	1.8
	THG73	100

The THG73 suppressor fared significantly better in the reacylation test than did THA73, which showed a higher degree of readthrough and, consequently, a much poorer efficiency ratio. Of the two positions, suppression at α180 is the more sensitive reacylation test because this site is more tolerant of amino acid substitutions and as such is more likely to produce functional nAChR as a result of readthrough. The efficiency ratio of THG73 was much better than MN3 at position α180, though it was somewhat lower at the α198 site.

In other experiments, THG73 and MN3 were used to introduce tyrosine or leucine residues at two different positions in the nAChR α-subunit and at one site in each of the other three subunits: α198, α93, β262, γ260, and δ265. In each case the current measured following THG73 injection was at least five times greater than MN3 (Table 1), showing that more protein is produced using this tRNA. Several other factors were taken into account in this study that are specific to the particular case of nAChr and will not be discussed here. The advantages of this suppressor tRNA for introducing noncoded amino acids in the oocyte system are clear; its utility in an *E. coli* transcription-translation system has also been demonstrated (see Sect. 4.2).

4.2
Escherichia coli tRNA[Asn]-Derived Suppressor tRNA

Unsatisfied with the relatively low suppression efficiency of their yeast tRNA[Phe]-derived suppressor, particularly when charged with amino acids having highly polar sidechains, the Schultz group conducted a search for a better suppressor [33]. The *E. coli* ribosome was known to have a greatly reduced affinity for yeast tRNA[Phe] compared to the *E. coli* tRNA[Phe], which was suspected to cause a reduction in the efficiency of suppression. For this reason, the search was limited to *E. coli* tRNAs. A derivative of a tRNA that normally carries a polar amino acid could be expected to be more effective in introducing polar noncoded amino acids, and amber suppressors derived from tRNA[Asn] and tRNA[Asp] were chosen for further analysis. It had previously been shown by Kleina et al. that these suppressors have virtually no activity in vivo [34], presumably because they are no longer recognized by aminoacyl-tRNA synthetases.

The ability of both suppressor tRNAs to incorporate the nonpolar amino acid valine as well as the polar noncoded homoglutamate into two proteins was tested in *E. coli* cell-free transcription-translation systems [35]. The proteins T4 lysozyme and *E. coli* chorismate mutase contained TAG codons at positions 84 and 88, respectively. For comparison, the Dougherty lab's amber suppressor tRNA^{Gln} [36] and the Chamberlin lab's tRNA^{Gly}-derived amber suppressors were also tested. The results are shown in Figs. 5 and 6. Readthrough levels were low for all of the suppressor tRNAs, confirming that they were not recognized by endogenous aminoacyl synthetases.

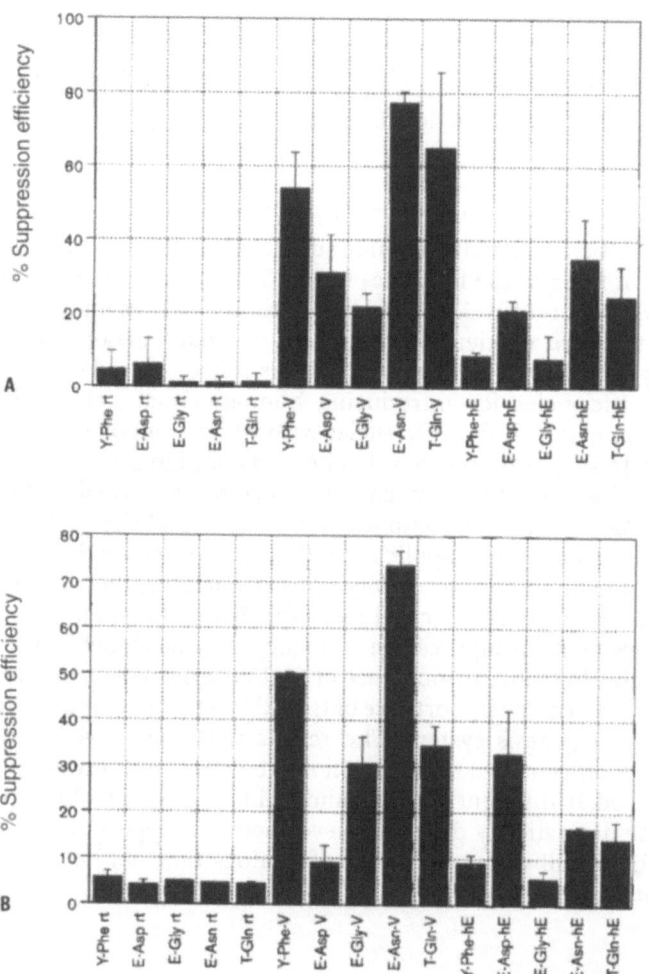

Fig. 5. Comparison of suppression efficiencies of five tRNAs **A** in T4 lysozyme at site 82, and **B** in chorismate mutase at site 88. Suppression efficiencies are defined as the amount of full-length protein divided by the sum of the full-length and truncated protein produced in each reaction. The suppression efficiencies shown represent the average of two trials. The tRNAs are identified below each bar: *Y* yeast, *E E. coli; T Tetrahymena; rt* readthrough (un-acylated tRNA); *V* acylated with valine; *hE* acylated with homoglutamate. Reprinted with permission [33]

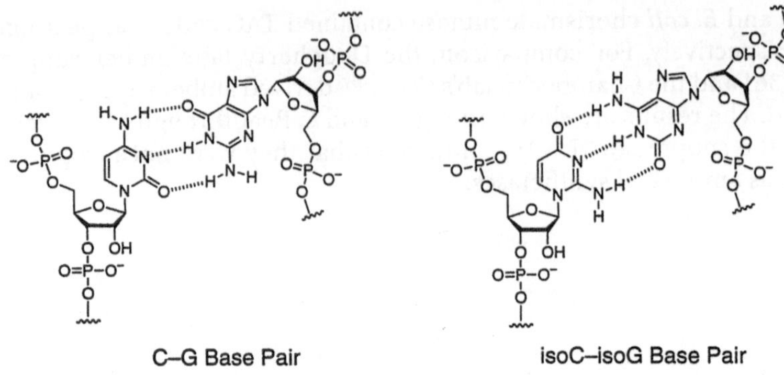

C–G Base Pair   isoC–isoG Base Pair

Fig. 6. Comparison of C–G and isoC–isoG base pairs

Overall, the *E. coli* tRNA^{Asn}- and *T. thermophila* tRNA^{Asn}-derived suppressors were the most efficient, having the same suppression efficiencies, within experimental error, at three of the four sites tested. The hypothesis that a suppressor tRNA patterned after a tRNA species that naturally carries a polar amino acid would lead to more efficient introduction of polar analogs was therefore supported by this study. The three such tRNAs tested showed decidedly higher suppression efficiencies for introducing homoglutamate, although the best results still cannot match the efficiencies with which nonpolar amino acids are incorporated. In support of the idea that nonpolar amino acids are simply better substrates for the biosynthetic machinery, these three showed higher efficiencies for introduction of valine, with the sole exception of tRNA$^{Asp, CUA}$ in chorismate mutase. In a subsequent application of the suppressor tRNA$^{Asp, CUA}$ to introduce S-methylmethionine into staphylococcal nuclease, the suppression efficiency was 15–20%, which is a good result for this analog [37].

The improvements in suppression efficiency that the Dougherty and Schultz groups were able to realize through use of tRNAs derived from different natural tRNA species confirm the importance of the tRNA sequence and structure in the suppression mutagenesis system. The results must be viewed with caution, however, as the efficiency of suppression is subject to other variables and is not fully understood. In different proteins, and at different sites within a protein, the results can be dramatically different. Nevertheless, it appears that use of these improved tRNAs will widen the range of analogs that can be introduced and increase protein yields.

5
Competition with Release Factors

During the normal process of termination of translation, stop codons are recognized by protein release factors (RF). Although the details of the process are not fully understood, it is believed that when a termination codon reaches the ribosomal A-site, the RF associates with the ribosomal–mRNA complex, inducing the peptidyl-transferase center to hydrolyze the ester bond of the peptidyl-tRNA in the P-site, releasing the polypeptide chain [38, 39]. During protein expression there is a direct competition between the suppressor tRNA and the release factor for binding to the A-site when the ribosome encounters a termination codon [16].

This competition was quantified by directly comparing the efficiency of incorporation of iodotyrosine into a hexadecapeptide by two different methods [17]. The first method relied on suppression of a termination codon while the second relied on expansion of the genetic code by creating a 65th codon-anticodon pair with the nonstandard nucleoside bases *iso*C and *iso*G, which have a specific, "nonnatural" hydrogen-bonding pattern (Fig. 6) [40]. Two mRNAs coding for hexadecapeptides were synthesized; the first contained an amber (UAG) termination codon at position 9 and the second contained the (*iso*C)AG codon at that position. These mRNAs were added to a rabbit reticulosyte lysate along with a synthetic iodotyrosyl-tRNA (iTyr) containing either a CUA or a CU(*iso*G) anticodon. In each system three products were possible: (a) a 16-mer if the iTyr was incorporated; (b) an 8-mer if the iTyr was not incorporated and peptide synthesis was terminated; and (c) a 12-mer if the iTyr was not incorporated but the mRNA frame shifted and peptide synthesis was terminated at the next stop codon.

The systems were evaluated by following the incorporation of iTyr at position 9. The system containing the amber suppressor iTyr-tRNA$_{CUA}$ yielded the 16-mer peptide containing iTyr with 67% efficiency, while the 8-mer termination product was produced in high yield in the absence of the charged suppressor tRNA. In the system containing iTyr-tRNA$_{CU(isoG)}$ the 16-mer containing iTyr was produced with 91% efficiency. In this case, when no charged tRNA$_{CU(isoG)}$ was present little 8-mer termination product was produced, but the 12-mer frame-shift product was produced in high yield.

These experiments make it clear that removing competition with release factors leads to more efficient incorporation of the desired amino acid. Unfortunately, the technology to incorporate nonstandard nucleotides into mRNAs coding for full-length proteins is not yet available. Alternatives that have been tested include using (i) a 4-nucleotide codon-anticodon pair, dubbed frameshift suppression (Sect. 6.1), (ii) a rare codon, and (iii) cell-free extracts from organisms that are either deficient in a release factor (Sect. 5.1) or unable to translate one or more codons (Sect. 6.2).

5.1
Release Factor Deficient Lysates

Competition with release factors can be reduced by deactivating a release factor in a cell-free lysate prepared from a prokaryote. This option is not available in rabbit reticulyte lysate since mammals have only two release factors (RFs), termed eukaryotic release factors 1 and 3 (eRF1 and eRF3). In eukaryotes, eRF1 recognizes all three stop codons and eRF3 stimulates eRF1 activity in the presence of GTP [41]. Deactivation of either RF would effectively deactivate all three stop codons, and at least one termination codon must function properly for the suppression method to be useful.

Deactivation of an RF is feasible, however, in prokaryotes. In *E. coli*, three RFs are involved in the termination process, and there is a degree of codon specificity for two of them. RF1 recognizes the UAG and UAA stop codons, while RF2 recognizes the UGA and UAA stop codons [16, 42]. RF3, which is not codon specific, stimulates the activity of the other two. Thus, if RF1 is deactivated, UAA and UGA remain functional termination codons, but UAG should no longer signal for termination of protein synthesis. To test this hypothesis, the Chamberlin lab investigated a mutant strain of *E. coli* that produces a faulty RF1 [29].

It has been shown previously that *E. coli* strain US477 increases the efficiency of amber tRNA suppressors in vivo at permissive growth temperatures and cannot grow at 43 °C [34, 43]. The mutation had been mapped to the same location on the chromosome as the RF1 gene [44], and it was assumed RF1 contained a mutation causing the temperature sensitivity. A related strain contains a point mutation within a region of RF1 that is proposed to be involved in ribosome binding [45, 46], further suggesting that such an RF1 mutant might be thermally deactivated.

Two S30 lysates were prepared from a single growth of US477 cells, one of which was heat-shocked at 43 °C for 15 min and designated US477-hs lysate. Lysate was also prepared from the parent strain, US475, which does not contain the temperature-sensitive RF. Transcription reactions of β-galactosidase from plasmid containing a TAG codon at position 7 were carried out using suppressor tRNA$^{Gly, CUA}$ acylated with leucine. The initial, unoptimized suppression assays looked promising. The US477 lysate only produced approximately 40 % as much wild-type β-galactosidase as the US475 lysate; however, twice as much β-galactosidase was produced in the US477-hs suppression reaction as compared to the US475 suppression reaction, suggesting that suppression yields might be substantially increased using lysates with nonfunctional RFs.

A second set of suppression reactions was run with five different amino acids in each of the three lysates (US475, US477 and US477-hs), but the results were equivocal. Protein yields and suppression efficiencies [47] were essentially the same in all three lysates for each of the amino acids except phenylalanine. For phenylalanine, suppression efficiency was 17 % in US475, 20 % in US477 and 29 % in US477-hs. More recently, the Huber lab has shown that the system does result in increases of 10 % compared with standard lysates (personal communication).

6
Alternative Methodologies

6.1
Frame-Shift Suppression Mutagenesis

In an effort to reduce the competition encountered with naturally occurring tRNAs, even when rare codons are used, Hardesty and co-workers have devised a strategy based on 4-nucleotide codon-anticodon pairs [48]. An extra thymidine was inserted either 5' or 3' to the rare arginine codon AGG to yield TAGG and AGGT codons. These 4-nucleotide codons were designed to insert a stop codon to facilitate translational termination when the 4-nucleotide codon was not translated by the corresponding synthetic tRNA; suppression of the 4-base codon was required for in-phase translation to continue. Synthetic tRNAs derived from *E. coli* tRNA^{Ala,UGC} containing the corresponding anticodons were prepared and aminoacylated with alanine enzymatically in the transcription/translation system used to synthesize the protein.

To test the technique, TAG, AGG, TAGG and AGGT were each introduced as the codon for position 75 of dihydrofolate reductase (DHFR), a well-characterized protein whose properties were known to be essentially unaffected by inserting alanine at this site. The protein was expressed in an *E. coli* S30 system, and in all cases except that of the AGG mutant, full-length, enzymatically active DHFR was produced only upon addition of the synthetic tRNA, with production being more efficient using the AGGT codon than the TAGG codon. In the case of the AGG mutant, the amount of protein synthesized equaled that produced from the wild-type gene. It was shown that the products were a heterogeneous mixture resulting from competition with the natural tRNA^{Arg,CCU}. While the Hardesty group introduced only alanine, the system is clearly applicable to the incorporation of noncoded amino acids and has subsequently been used for that purpose.

The Sisido laboratory has introduced amino acids containing *p*-nitrophenyl, 2-naphthyl, *p*-(phenylazo)phenyl and 2-anthryl groups into streptavidin using 4-base codon-anticodon pairs [49]. The four AGGN-NCCU pairs were tested, once again relying on inefficient translation of AGG to favor frame-shift suppression. The four codons were introduced into streptavidin at position 83 such that full-length protein would be produced only if the frame-shift occurred; otherwise, an in-frame stop codon is encountered that results in production of truncated protein. Semi-synthetic tRNAs based on yeast tRNA^{Phe} were produced in accordance with the methodology described in Sect. 2. Full-length protein was produced in *E. coli* S30 lysate reactions with each of these codons only when the corresponding charged tRNA$_{NCCU}$ was included. A control reaction containing uncharged tRNA$_{ACCU}$ did not produce streptavidin, indicating that the tRNA is not aminoacylated by the endogenous tRNA synthetases. The suppression efficiency was estimated to be approximately 20% [47] for *p*-nitrophenylalanine based upon an estimated yield of 5 μg ml^{-1}.

The Sisido group recently reported the introduction of *p*-nitrophenylalanine (ntrPhe) into streptavidin in this manner for the study of site-to-site photoinduced electron transfer in proteins [50]. The ntrPhe serves as an electron acceptor and the electron donor is a pyrenyl group which works as a photosensitizer and is introduced by linking L-1-pyrenylalanine to biotin (Fig. 7), a substrate for which streptavidin has extremely high affinity ($\sim 10^{15}$ M^{-1}). The ntrPhe was incorporated at 22 sites, and dot blot analysis using biotin-linked alkaline phosphatase demonstrated that 14 of the mutants retained the ability to bind streptavidin. These mutants were used to study the electron transfer rates as a function of the edge-to-edge distances between the pyrenyl and ntrPhe groups.

One advantage of the frame-shift suppression system is that it offers the potential to site-specifically introduce more than one noncoded amino acid into a protein if the yields can be increased. Kanda and co-workers have attempted

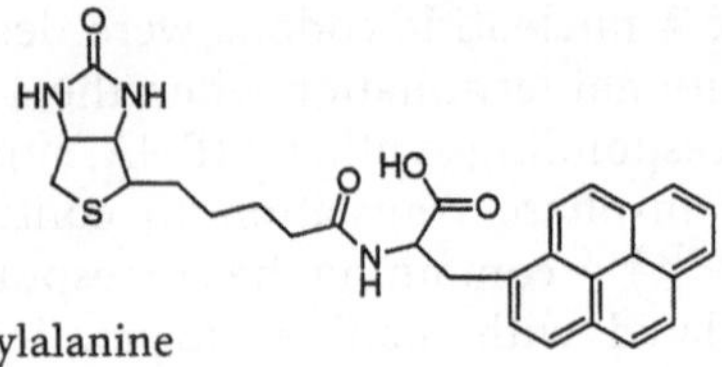

Fig. 7. Biotin-linked pyrenylalanine

to address the low yields by providing a way to remove endogenous tRNAs from S30 extracts [51] which would eliminate competition from natural tRNA$^{Arg, AGG}$. This is accomplished by incubating the S30 extract with resin-bound RNase A to digest the endogenous tRNAs and then filtering out the resin. The reactions are then supplied with tRNA mixtures that do not include tRNA$^{Arg, AGG}$. After some optimization, protein synthesis in the treated lysate was quantified by [^{14}C]arginine incorporation in three reactions: one to which no mRNA or tRNA was added, one to which only a tRNA mixture was added, and a third supplied with both mRNA and mixture of tRNAs. When a tRNA mixture was added, RNase inhibitor was also included at a concentration of 3200 units ml^{-1}. No protein was produced unless both mRNA and the tRNA mixture were added, and the activity in this case was close to that in the untreated extract. While they did not report testing protein production in the treated extract with a frame-shift suppressor, their results should increase yields.

6.2
Unassigned Codons in *Micrococcus luteus*

Another alternative to suppression of stop codons with the possibility of incorporating multiple noncoded amino acids into proteins has been demonstrated by Oliver and Kowal [52]. They prepared S30 extracts from *Micrococcus luteus*, a eubacterium with a high genomic GC content that is believed to be incapable of translating as many as six codons; a survey of codon usage in a sample of both highly and weakly expressed genes in this species showed that six codons were not used [53]. Further investigation demonstrated that the complementary tRNAs were also absent [54], making this an excellent candidate system for potentially efficient introduction of noncoded amino acids without competition from either endogenous tRNAs or termination factors.

Building on earlier work of Osawa and co-workers [55], Oliver and Kowal [52] tested the feasibility of introducing a noncoded amino acid at an unassigned codon in *M. luteus*. DNA templates were prepared which coded for 19-mer polypeptides containing either the unassigned codon AGA(Arg) or the termination codon TAG at position 13 under the control of a T7 RNA polymerase promoter. The corresponding tRNAs, produced as described in Sect. 2, were based on tRNA^{Ala} and acylated with phenylalanine. The tRNA was modified to prevent recognition by the alanine aminoacyl-tRNA synthetase and to increase translational efficiency.

When the DNA templates were added to *M. luteus* transcription-translation reactions containing the corresponding charged suppressor tRNAs, full-length peptides were produced, and they were not produced in control reactions containing uncharged suppressor tRNAs or to which no suppressor tRNA was added. Truncated peptides were not detected in the latter two cases, but the authors suggest this is due to low yields or degradation of the truncated peptides. Suppression efficiencies were determined as a percentage of peptide translated from the AGA-codon template in an *M. luteus* extract with the designer tRNA as compared to the amount produced when the extract was supplied with an *E. coli* tRNA$^{Arg, AGA}$. Incorporation efficiencies were 72% for the unassigned AGA codon and 53% for the TAG codon, indicating an advantage that is presumably due to the absence of competition from release factors.

While this methodology may not prove to be as generally applicable as suppression of stop codons, which can be accomplished in a variety of cell-free systems, it does, as its authors suggest, offer the possibility of introducing more than one type of noncoded amino acid into a protein and may work at sites where stop codons cannot be suppressed.

7
Applications of Site-Directed Suppression Mutagenesis

The ability to biosynthetically incorporate noncoded amino acids into proteins site-specifically has facilitated studies not previously possible. These include studies of protein stability, the initiation of protein translation, electron transfer, protein–protein and protein–membrane interactions, reversal of enzyme substrate specificity, and structure-function relationships, among others. A growing number of research labs have begun to report applications of this technique. A brief look at some recent applications of the suppression mutagenesis technique follows.

Hecht and co-workers have reported several interesting applications of this methodology [56]. The thermolabile firefly luciferase from *Luciola mingrelica* was used to investigate the structural basis for changes in the wavelength of emitted light [56a and references therein]. It is believed that the enzyme-bound substrate luciferin is converted to an oxyluciferin enolate dianion in the excited state, and that relaxation to the ground state produces bioluminescence. A single residue, serine 286, was replaced with a series of coded and noncoded amino acids. The natural amino acids tyrosine, lysine, leucine, glutamine, and serine (the wild-type residue) were introduced in vivo. For each of these luciferases, the emission maximum, pH response and thermostability were altered from the wild type, with the greatest changes seen upon introduction of the hydrophobic residue leucine. The in vitro introduction of the noncoded amino acids O-glucosylated serine, serine methylenephosphonate, tyrosine phosphate and tyrosine methylenephosphonate produced some surprising results. The luciferases containing the glucosylated and phosphonate derivatives of serine did not exhibit any change in λ_{max}. On the other hand, the emission wavelengths of luciferase containing a phosphorylated tyrosine and a nonhydrolyzable phosphonate analog of tyrosine are shifted substantially from each other as well as from the tyrosine-containing luciferase. The substituted amino acid is not directly involved in conversion of luciferin to oxyluciferin with the emission of light and the authors suggest that the differences in emitted light stem from differences in pK_a values for these residues. Based on this result, they caution against the frequent assumption of equivalence between phosphorylated amino acids and their phosphonate analogs.

In a collaboration between the Abelson and Hecht labs [56b], a series of noncoded amino acids were introduced into dihydrofolate reductase (DHFR) to probe substrate binding and the requirement of an aspartic acid residue for catalytic competence. When aspartic acid analogs mono- or disubstituted at the β-carbon were substituted for the active site aspartic acid residue, the mutant DHFRs were still able to catalyze the NADPH-dependent reduction of dihydrofolate to tetrahydrofolate at 74–86% of the wild-type rate. While hydride transfer from NADPH is not the rate-limiting step for the wild-type enzyme at physiological pH, a kinetic isotope experiment with NADPD indicated that hydride transfer had likely become the rate-limiting step for the mutant containing the β,β-dimethylaspartic acid.

In another interesting application, the introduction of backbone mutations into proteins has been used by the Schultz lab to evaluate the strength of main-chain hydrogen-bonding interactions in proteins [57]. (An earlier study had considered the strength of hydrogen bonds between side chains [58].) In one set of experiments, α-hydroxy acids were used to introduce an ester linkage at different locations in an α-helix of T4 lysozyme [57b], while in the other the ester linkages replaced amide bonds in an anti-parallel β-sheet of staphylococcal nuclease [57a] (ester incorporation is easily verified by selective alkaline hydrolysis). The ester carbonyl group is a weaker hydrogen-bond acceptor than the amide carbonyl, but is structurally similar. In the case of the β-sheet amide to ester substitutions, mutation sites were chosen so that the ester oxygen was solvent exposed. It was found that the protein was destabilized by 1.5–2.5 kcal mol^{-1} based on apparent equilibrium constants determined from guanidine hydrochloride denaturation curves. In the T4 lysozyme α-helix, esters were introduced at the N- and C-terminal positions and at the middle of the helix; one hydrogen-bonding interaction was altered at each terminus, and two were altered for the residue at the center of the helix. Thermal denaturation measurements indicated decreases in stability for the ester-containing helices of 0.9 kcal mol^{-1} with the ester at the N-terminus, 0.7 kcal mol^{-1} at the C-terminus, and 1.7 at the central position. A useful discussion of the factors contributing to these free-energy differences is included.

Structural studies of integral membrane proteins and of protein interactions with membranes or membrane-bound proteins are benefiting particularly from the versatility that suppression mutagenesis offers. High et al. have used site-specific introduction of a photoactivated cross-linking residue to facilitate study of contacts between the signal sequence of a secretory protein and the protein subunits of the protein conducting channel in the endoplasmic reticulum (ER) membrane [59]. Subsequent use of this cross-linking agent in conjunction with a technique to produce transmembrane-arrested intermediates has provided additional information about such contacts as well as providing evidence of cross-linking to membrane lipids [60]. They concluded from these experiments that the protein-conducting channel opens toward the lipid bilayer and that the signal sequence is positioned in the channel so that it contacts the lipid bilayer as well as channel proteins. The more hydrophilic residues of the polypeptide that follow the signal sequence appear to contact only the proteins of the channel as they are translocated across the membrane.

In another application of this methodology to an integral membrane protein, Rothschild and co-workers have site-specifically incorporated isotopically labeled amino acids into bacteriorhodopsin to facilitate analysis of structural changes by Fourier transform infrared (FTIR) difference spectroscopy [61]. Rothschild relied on a slightly modified protocol in which an E. coli tyrosyl amber suppressor tRNA is enzymatically aminoacylated with [^{2}H]-tyrosine by E. coli tyrosyl tRNA synthetase in vitro. By introducing the labeled tyrosine at specific sites, they were able to demonstrate that just one of the 11 tyrosines in bacteriorhodopsin undergoes important structural changes during the bacteriorhodopsin photocycle [61c].

The highly successful extension of site-directed suppression mutagenesis to a Xenopus oocyte system by the Dougherty and Lester labs discussed above has permitted structural, functional and kinetic studies of ion channels in intact cells [30–32, 62]. Recently, the introduction of residues that allow photochemical proteolysis and photodeprotection of protein side chains has been applied to the study of ion channels. Miller et al. introduced o-nitrobenzyl-protected tyrosines at three positions in the α-subunit of nAChR [62 c]. The o-nitrobenzyl group was then removed photolytically to produce the wild-type residue during the course of a voltage-clamp study, so that time-resolved measurements could be made.

England et al. [62b] incorporated (2-nitrophenyl)glycine (Npg) into the nAChR and the Drosophila Shaker B K$^+$ ion channel. The latter channel was used as a test case because deletion of certain residues (Δ6–46) is known to convert the channel to Shaker-IR, which does not exhibit the same inactivation characteristics as the wild-type channel. The mutated channels did show wild-type activity prior to photolysis, and afterwards exhibited the same characteristics as systems containing a mixture of the wild-type and Shaker-IR channels. The technique was then used to probe whether a highly conserved disulfide loop, already known to be required for proper folding and receptor assembly, has any functional role. Npg was incorporated in the disulfide loop in the N-terminal extracellular domain of the α-subunit and, prior to photolysis, these channels also displayed wild-type characteristics. It was proposed that cleavage of the peptide backbone in this loop would result in nothing more than local perturbations of structure since the disulfide bond would be left intact. Backbone cleavage of a critical residue in a trans-membrane region was shown to reduce whole-cell current by half without affecting surface binding of α-bungarotoxin, which requires a properly folded receptor. Irradiation of the channels with Npg in the disulfide loop led to near total loss of function, indicating that the loop may have a functional role, though it remains to be demonstrated that this cleavage did not result in global structural changes that could also account for loss of activity.

The oocyte system has also been utilized by Turcatti et al. to incorporate a fluorescent amino acid into the tachykinin neurokinin-2 receptor (NK2), allowing microspectrofluorimetry of the protein within the native membrane environment [63]. The intermolecular distances between the site-specifically incorporated fluorescent residue and a fluorescent NK2 antagonist were measured by fluorescence resonance energy transfer (FRET). Based on the FRET data, the authors proposed a model for NK2 ligand–receptor interactions.

8
Conclusions

Site-directed suppression mutagenesis can expand the range of questions that can be asked about proteins as well as increasing the specificity of these questions. However, the methodology in its current form suffers from several limitations, some of them serious. Among these are its labor-intensive nature, the small amounts of protein produced, the restrictions imposed by the biosynthetic machinery on the types of amino acids that can be incorporated, and context effects that currently cannot be anticipated and which unpredictably preclude the introduction of amino acids at certain sites.

At this stage, it is clear that ribosomal protein synthesis will not allow the incorporation of D-amino acids or those that are too bulky, and that only α-amino and α-hydroxy acids can be introduced; those with extended backbones such as γ-amino acids cannot be introduced in this manner. Also, in spite of the development of improved suppressor tRNAs, the incorporation of small, highly polar amino acids remains difficult.

Low protein yields may not be a problem if, for example, enzymatic activity is being measured with a sensitive assay. If greater amounts of protein are needed, such as for structural studies requiring NMR spectroscopy or X-ray crystallography, reaction volumes can be increased. While reaction volumes are typically less than 250 µl, 5 and even 10 ml reactions have been reported. However, although this methodology is amenable to scaling up, the cost quickly becomes prohibitive unless the lysates are prepared and the enzymes for tRNA ligation and mRNA production are purified in-house.

Even with these limitations, site-directed suppression mutagenesis has opened new possibilities and made a wide range of experiments possible. More laboratories are utilizing the methodology and are working to improve the techniques. It is likely that such improvements will facilitate the application of suppression mutagenesis as a useful tool in the study of proteins.

9
References

1. Winter G, Fersht AR, Wilkinson AJ, Zoller M, Smith M (1982) Nature 299:756
2. Ulmer KM (1983) Science 219:666
3. Glover DM, Hames BD (eds) (1995) DNA cloning: a practical approach, 2nd edn. IRL Press, New York
4. (a) Kent SB, Baca M, Elder J, Miller M, Milton R, Milton S, Rao JMK, Schnolzer M (1995) Adv Exp Med Biol 362:425; (b) Nyfeler R (1994) Methods Molec Biol 35:303; (c) Muir TW, Kent SBH (1993) Curr Opin Biotechnol 4:420; (d) Humphries J, Offord RE, Smith RAG (1991) Curr Opin Biotechnol 2:539

5. (a) Offord RE (1987) Protein Eng 1:151; (b) Chaiken IM (1981) Crit Rev Biochem 11:255
6. (a) Gittel C, Schmidtchen FP (1995) Bioconjugate Chem 6:70; (b) Merkler DJ (1994) Enzyme Microb Technol 16:450; (c) Manning JM (1994) Methods Enzymol 231:225; (d) Matthews KS, Chakerian AE, Gardner JA (1991) Methods Enzymol 208:468; (e) Means GE, Feeney RE (1990) Bioconjugate Chem 1:2; (f) Means GE, Feeney RE (1971) Chemical modification of proteins. Holden-Day, San Francisco
7. (a) Ross JBA, Szabo AG, Hogue CWV (1997) Methods Enzymol 278:151; (b) Wilson MJ, Hatfield DL (1984) Biochim Biophys Acta 781:205; (c) Hortin G, Boime I (1983) Methods Enzymol 96:777
8. (a) Do H, Falcone D, Lin J, Andrews DW, Johnson AE (1996) Cell 85:369; (b) Mothes W, Prehn S, Rapoport TA (1994) EMBO J 13:3973; (c) Johnson AE (1993) Trends Biochem Sci 18:456; (d) High S, Andersen SSL, Görlich D, Hartmann E, Prehn S, Rapoport TA, Dobberstein B (1993) J Cell Biol 121:743
9. (a) Bain JD, Glabe, CG, Dix, TA and Chamberlin, AR (1989) J Am Chem Soc 111:8013; (b) Bain JD, Diala ES, Glabe CG, Wacker DA, Lyttle MH, Dix TA, Chamberlin, AR (1991) Biochemistry 30:5411; (c) Bain JD, Wacker DA, Kuo EE, Lyttle MH, Chamberlin AR (1991) J Org Chem 56:4615; (d) Noren CJ, Anthony-Cahill SJ, Griffith MC, Schultz PG (1989) Science 244:182; (d) Ellman J, Mendel D, Anthony-Cahill S, Noren CJ, Schultz PG (1991) Methods Enzymol 202:301
10. (a) Söll D (1988) Nature 331:662; (b) Murgola EJ (1995) Translational suppressors: when two wrongs DO make a right. In: Söll D, RajBhandary U (eds) tRNA: Structure, biosynthesis and function. ASM Press, Washington, DC, p 491
11. (a) Hecht SM, Alford BL, Kuroda, Y, Kitano, SJ (1978) J Biol Chem 253:4517; (b) Pezzuto JM, Hecht SM (1980) J Biol Chem 255:865; (c) Heckler TG, Zama Y, Naka T, Hecht SM (1983) J Biol Chem 258:4492; (d) Heckler TG, Chang LH, Zama Y, Naka T, Chorghade MS, Hecht SM (1984) Biochemistry 23:1468; (e) Heckler TG, Chang LH, Zama Y, Naka T, Hecht SM (1984) Tetrahedron 40:87; (f) Payne RC, Nichols BP, Hecht SM (1987) Biochemistry 26:3197; (g) Heckler TG, Roesser JR, Xu C, Chang P-I, Hecht SM (1988) Biochemistry 27:7254; (h) Roesser JR, Xu C, Payne RC, Surratt CK, Hecht SM (1989) Biochemistry 28:5185
12. Baldini G, Martoglio B, Schachenmann A, Zugliani C, Brunner J (1988) Biochemistry 27:7951
13. Englisch U, Gauss D, Freist W, Englisch S, Sternbach H, von der Haar F (1985) Angew Chem Int Ed Engl 24:1015
14. (a) Bain JD, Wacker DA, Kuo EE, Lyttle MH, Chamberlin AR (1991) J Org Chem 56:4608; (b) Lyttle MH, Wright PB, Sinha ND, Bain JD, Chamberlin AR (1991) J Org Chem 56:4615; (c) Noren CJ, Anthony-Cahill JA, Suich DJ, Noren KA, Griffith MC, Schultz PG (1991) Nucleic Acids Res 18:83
15. (a) Steward LE, Chamberlin AR (1998) Methods Molec Biol 77:325; (b) Ellman J, Mendel D, Anthony-Cahill S, Noren CJ, Schultz PG (1991) Methods Enzymol 202:301
16. Caskey CT, Forrester WC, Tate W (1984) Peptide chain termination. In: Clark BFC, Petersen HU (eds) Gene expression: the translational step and its control. Munksgaard, Copenhagen
17. Bain JD, Switzer C, Chamberlin AR, Benner SA (1992) Nature 356:537
18. Hatfield GW, Gurman GA (1992) Codon pair utilization bias in bacteria, yeast and mammals. In: Hatfield DL, Lee BJ, Pirtle RM (eds) Transfer RNA in protein synthesis. CRC Press, Boca Raton, chap 7
19. Noren CJ, Anthony-Cahill S, Suich DJ, Noren KA, Griffith MC, Schultz PG (1989) Nucl Acids Res 18:83
20. Bain JD, Wacker DA, Kuo EE, Chamberlin AR (1991) Tetrahedron 47:2389
21. Unless otherwise noted, suppression efficiencies were calculated by dividing the amount of full length mutant protein by the combined total of full length and truncated protein
22. Hecht SM (1992) Acc Chem Res 25:545
23. (a) Mendel D, Cornish VW, Schultz PG (1995) Annu Rev Biophys Biomol Struct 24:435; (b) Cornish VW, Mendel D, Schultz PG (1995) Angew Chem Int Ed Engl 34:621

24. (a) Fahnestock S, Neuman H, Sashova V, Rich A (1970) Biochemistry 9:2477; (b) Scolnik E, Milman G, Rosman M, Caskey T (1970) Nature 225:152; (c) Fahnestock S, Rich A (1971) Science 173:340
25. Pallanch L, Pak M, Schulman LH (1995) tRNA discrimination in aminoacylation. In: Soll D, RajBhandary U (eds) tRNA: structure, biosynthesis, and function. ASM Press, Washington, DC, chap 18
26. Fersht AR, Dingwall C (1979) Biochemistry 18:2627
27. The last six bases of the promoter, the +1 to +6 region, are transcribed and have been shown to affect transcription yields
28. (a) Rosa MD (1979) Cell 16:815; (b) Milligan JF, Groebe DR, Witherell GW, Uhlenbeck OC (1987) Nuc Acids Res 15:8783
29. Steward LE (1996) PhD thesis, University of California, Irvine
30. Nowak MW, Kearney PC, Sampson JR, Saks ME, Labarca CG, Silverman SK, Zhong W, Thorson J, Abelson JN, Davidson N, Schultz PG, Dougherty DA, Lester HA (1995) Science 268:439
31. (a) Kearney PC, Nowak MW, Zhong W, Silverman SK, Lester HA, Dougherty DA (1996) Molecular Pharmacology 50:1401; (b) Kearney PC, Zhang H, Zhong W, Dougherty DA, Lester HA (1996) Neuron 17:1221
32. Saks M, Sampson JR, Nowak MW, Kearney PC, Du F, Abelson JN, Lester HA, Dougherty DA (1996) J Biol Chem 271:23169
33. Cload ST, Liu DR, Froland WA, Schultz PG (1996) Chem & Biol 3:1033
34. Kleina LG, Masson JM, Normanly J, Abelson J, Miller JH (1990) J Mol Biol 213:705
35. A suppressor tRNA corresponding to the termination codon TGA was also tested, but readthrough levels exceeding 90% made it unsuited for this purpose
36. Chg to a cross reference to Dougherty's tRNA paper (1996)
37. Ting, AY, Shin I, Lucero C, Schultz PG (1998) J Am Chem Soc 120:7135
38. Spirin AS (1986) Ribosome structure and protein biosynthesis. Benjamin/Cummings Publishing, Menlo Park
39. Tate WP, Brown CM (1992) Biochemistry 31:2443
40. Benner SA (1994) Trends Biotechnol 12:158
41. (a) Zhouravleva G, Frolova L, Le Goff X, Le Guellec R, Inge-Vechtomov S, Kesselev L, Philippe M (1995) EMBO J 14:4065; (b) Frolova L, Le Goff X, Zhouravleva G, Davydova E, Philippe M, Kisselev L (1996) RNA 2:334
42. Caskey CT (1998) Trends Biochem Sci 5:234
43. Rydén SM, Isaksson LA (1984) Mol Gen Genet 193:38; Rydén M, Murphy J, Martin R, Isaksson L, Gallant J (1986) J Bacteriol 168:1066
44. Rydén M, Murphy J, Martin R, Isaksson L, Gallant J (1986) J Bacteriol 168:1066
45. Pel HJ, Rep M, Grivell LA (1992) Nuc Acids Res 20:4423
46. Moffat JG, Donly BC, McCaughan KK, Tate WP (1993) J Biochem 213:749
47. Suppression efficiency was calculated in this instance as a percentage of the amount of wild-type protein produced in the same in vitro system
48. Ma C, Kudlicki W, Odom OW, Kramer G, Hardesty B (1993) Biochemistry 32:7939
49. Hohsaka T, Ashizuka Y, Murakami H, Sisido M (1996) J Am Chem Soc 118:9778
50. Murakami H, Hohsaka T, Ashizuka Y, Sisido M (1998) J Am Chem Soc 120:7520
51. Kanda T, Takai K, Yokoyama, Takaku H (1997) Nucleic Acids Symposium Series 37:319
52. Kowal AK, Oliver JS (1997) Nucl Acids Res 25:4685
53. Ohama T, Muto A, Osawa S (1990) Nucl Acids Res 18:1565
54. Kano A, Andachi Y, Ohama T, Osawa S (1991) J Mol Biol 221:387
55. Kano A, Ohama T, Abe R, Osawa S (1993) J Mol Biol 230:51
56. (a) Arslan T, Manaev SV, Mamaeva NV, Hecht SM (1997) J Am Chem Soc 119:10877; (b) Karginov VA, Mamaev SV, An H, Van Cleve MD, Hecht SM, Komatsoulis GA, Abelson JN (1997) J Am Chem Soc 119:8166; (c) Karginov VA, Mamaev SV, Hecht SM (1997) Nuc Acids Res 25:3912; (d) Killian JA, Van Cleve MD, Shayo YF, Hecht SM (1998) J Am Chem Soc 120:3032

57. (a) Chapman E, Thorson JS, Schultz PG (1997) J Am Chem Soc 119:7151; (b) Koh JT, Cornish VW, Schultz PG (1997) Biochemistry 36:11314
58. Thorson JS, Chapman E, Schultz PG (1995) J Am Chem Soc 117:9361
59. High S, Martoglio B, Gorlich D, Andersen SSL, Ashford AJ, Giner A, Hartmann E, Prehn S, Rapoport TA, Dobberstein B, Brunner J (1993) J Biol Chem 268:26745
60. Martoglio B, Hofmann MW, Brunner J, Dobberstein B (1995) Cell 81:207
61. (a) Sonar S, Lee CP, Coleman M, Patel N, Liu X, Marti T, Khorana HG, RajBhandary UL, Rothschild KJ (1994) Nature Struct Biol 1:512; (b) Ludlam CFC, Sonar S, Lee C-P, Coleman M, Herzfeld J, RajBhandary UL, Rothschild KJ (1995) Biochemistry 34:2; (c) Liu X-M, Sonar S, Lee C-P, Coleman M, RajBhandary UL, Rothschild KJ (1995) Biophys Chem 56:63
62. (a) Kearney PC, Zhang H, Zhong W, Dougherty DA, Lester HA (1996) Neuron 17:1221; (b) England PM, Lester HA, Davidson N, Dougherty DA (1997) Proc Natl Acad Sci USA 94:11025; (c) Miller JC, Silverman SK, England PM, Dougherty DA, Lester HA (1998) Neuron 20:619
63. Turcatti G, Nemeth K, Edgerton MD, Meseth U, Talabot F, Peitsch M, Knowles J, Vogel H, Chollet A (1996) J Biol Chem 271:19991
64. Steward LE, Collins CS, Gilmore MA, Carlson JE, Ross JBA, Chamberlin AR (1997) J Am Chem Soc 119:6
65. Park Y, Luo J, Schultz PG, Kirsch JF (1997) Biochemistry 36:10517
66. Murakami H, Hohsaka T, Ashizuka Y, Sisido M (1998) J Am Chem Soc 120:7520
67. Zhao Z, Liu X, Shi Z, Danley L, Huang B, Jiang R-T, Tsai M-D (1996) J Am Chem Soc 118:3535

Catalysis Based on Nucleic Acid Structures

Michael Famulok · Andreas Jenne

Institut für Biochemie der LMU München, Feodor-Lynen-Str. 25, D-81377 München, Germany.
E-mail: Famulok@lmb.uni-muenchen.de

Since the discovery that RNA molecules can possess catalytic activities, ribozymes have become a fascinating field both for academic researchers and the pharmaceutical industry. In this review, we emphasize the latest progress made in structure determination of ribozymes as well as the generation of DNA and RNA enzymes with novel catalytic properties by combinatorial approaches.

Keywords: Ribozymes, In vitro selection, Nucleic acid libraries, Metallo enzymes, Aptamers.

Topics in Current Chemistry, Vol. 202
© Springer-Verlag Berlin Heidelberg 1999

1
Introduction

For a long time it was thought that among all known biopolymers, proteins are the only ones that are capable of catalyzing chemical transformations. In 1982/1983, however, it was discovered that naturally occurring RNA sequences, the group I introns [1] and the catalytic RNA subunit of RNase P [2], are able to catalyze the hydrolysis and formation of phosphodiester bonds. Since then, a large number of natural ribozymes have been described. Remarkably, it has recently become evident that the peptidyl transferase activity of the ribosomal protein/RNA-complex which is responsible for protein synthesis in living organisms is carried out by the RNA rather than the protein portion [3]. While this discovery has widened the scope of natural RNA catalysis from phosphodiester chemistry to the formation of peptide bonds, another dimension was introduced into nucleic acid catalysis by applying the techniques of in vitro selection of combinatorial nucleic acid libraries to isolate non-natural ribozymes, deoxyribozymes, or chemically modified nucleic acids with novel catalytic properties. These techniques have led to nucleic acids that catalyze a broad range of chemical transformations, ranging from cleavage of carboxylic ester- [4] or amide bonds [5] to C–C-bond-forming reactions such as the Diels-Alder reaction [6] or the catalysis of σ-bond-rotation involved in isomerization reactions [7]. While these examples of nucleic acid catalysis have an enormous implication for the support of theories about the origin of life on our planet, there are also several impressive examples that show that ribozyme catalysis opens up the possibility for completely novel therapeutic approaches [8–14]. Furthermore, the combination of specific ligand-binding nucleic acid sequences, the so-called aptamers, with catalytic RNAs have allowed to use ribozyme catalysis in a controlled and designed fashion [15–18]. In this review we will summarize the state-of-the-art of nucleic acid catalysis.

2
Natural Ribozymes

Since the discovery of the first ribozyme, the self-splicing intron of the ribosomal large subunit RNA of the ciliate *Tetrahymena thermophila* [19], far more than 100 other sequences that belong to the family of group I introns have been identified. To these can be added a considerable number of group II introns [20, 21], which act by a different RNA-cleavage mechanism, a large variety of virus- or viroid-derived ribozymes, namely the hammerhead- [22], the hepatitis delta virus- [23], the hairpin-ribozyme [24, 25], the M1-RNA subunit of RNase P which is responsible for tRNA 5′-end maturation in prokaryotes [26], and the neurospora mitochondrial VS RNA [27]. All these ribozymes have been the subject of a number of excellent reviews [28] giving us the opportunity in this chapter to focus on the more recent developments.

2.1
Structure and Mechanism of Some Natural Ribozymes

In mid-1997 an international conference took place in Santa Cruz, USA, in which, for the first time, the exclusive topic was structural aspects of RNA molecules. A report covering this meeting contains an impressive graphic which shows the RNA structures, RNA/DNA complexes, and RNA/protein complexes contained in the brookhaven database as a function of the year of their publication [29]. Between 1988 and 1993 there were just 20. However, in 1996 alone no less than 41 structures appeared. These new dimensions were headed by the crystal structural elucidation of the first larger RNA molecule since the first crystal structure of tRNA in 1973 [30], the 48 nucleotide long hammerhead ribozyme (HHR) [31–33]. This landmark achievement was followed by a crystal structure analysis of the P4-P6-domain of a group I intron [34–36] and, more recently, a crystal structure of the hepatitis delta virus ribozyme [37].

All these studies established that the folding of RNA molecules is organized in a hierarchical manner. The assembly of complex RNA structures occurs in discrete transitions which build upon the folding of sub-systems that can then self-organize to even bigger and more complex structures [38–40].

2.1.1
The Hammerhead Ribozyme

In 1994 and 1995, two crystal structures of hammerhead ribozymes [31, 32] and a structural analysis based on fluorescence resonance energy transfer studies [41] were published. In case of the crystal structure analyses, both ribozyme variants contained certain modifications that had been introduced to avoid self-cleavage [31, 32]. In one case a DNA-analog of the substrate oligonucleotide was used [31], in the other case the all-RNA substrate contained a $2'-O-CH_3$ modification at the attacking 2'-OH group to avoid cleavage in the crystal [32]; for reviews see [8, 42, 43].

While these structures gave important insights into the HHR-structure, the question remained to what extent the folding of a catalytically active HHR would differ from these structures (Fig. 1). A third crystal structure, this time using an unmodified HHR which was crystallized in the absence of any divalent metal ions, turned out to be quite conclusive [33]. After addition of Mg^{2+} the ability of the HHR for self-cleavage in the crystal was restored. In this active conformation the ribozyme folded in the same way as did the $2'-O-CH_3$-modified HHR. When the Mg^{2+} was added at low pH the ribozyme could be obtained in a metal-bound uncleaved form. This structure was compared with a conformational intermediate which contained a second Mg^{2+}-ion bound at the cleavage site and which could be analyzed at an active pH in its uncleaved form after shock-freezing. The most notable differences between this and the inactive form were located at the active site. These structures allowed the identification of five metal binding centers. They were also used to predict binding pockets for metal ions in any RNA molecule by measuring the distribution of electronegativity potentials [44]. Further insight into the cleavage mechanism could be obtained

A

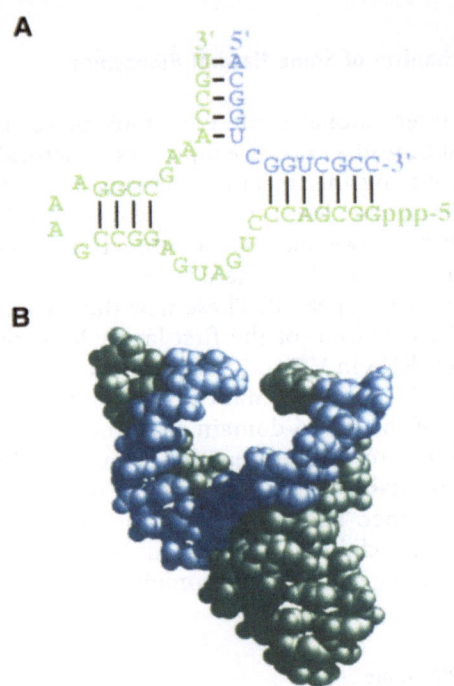

B

Fig. 1. The Hammerhead ribozyme. **A** Sequence and secondary structure. **B** Three dimensional structure of the HHR according to Scott et al. [33]. The substrate oligonucleotide (*blue*) is hybridized to the catalytic part (*cyan*)

by molecular dynamics simulations on the basis of the structure of the active ribozyme [45]. This analysis established that the two Mg^{2+}-ions in proximity to the cleavage site are bridged by an OH^--ion which activates the attacking 2'-OH group after the ribose at the phosphodiester group to be cleaved from C3'-*endo* into the C2'-*endo* conformation. Taira and his colleagues also significantly contributed to the understanding of the cleavage mechanism by the HHR. They provided evidence that the departure of the 5'-oxygen is the rate-limiting step in HHR-cleavage [46].

2.1.2
The Group I Introns

A landmark achievement in RNA structure determination was the solution of the crystal structure of the 160 nucleotide long P4-P6 domain of the *Tetrahymena* group I intron [19, 34, 35, 39]. The P4-P6-domain folds into a compact structure with a sharp turn that is stabilized by tight packing of the helices. This newly discovered structural element was designated as the "ribose-zipper" because of the hydrogen bonds between ribose residues of the helices that participate in the structure. In addition, stabilization of RNA folds in P4-P6 occurs mainly via tertiary interactions of the eleven base long 'tetraloop receptor' with certain extra-stable GAAA hairpin loops that also contribute to the stability of the motif. In most natural RNAs, interactions of that sort play an important role in the stabilization of RNA tertiary structures [47], e.g. in the group II intron [48–50].

The crystal structure of the P4-P6 domain (Fig. 2) disclosed another novel structural motif [51], the "adenosine platform" [52]. This motif is unusual because it contains two consecutive adenosine residues arranged in the same plain. Adenosine-platforms appear three times in the crystal structure and are part of

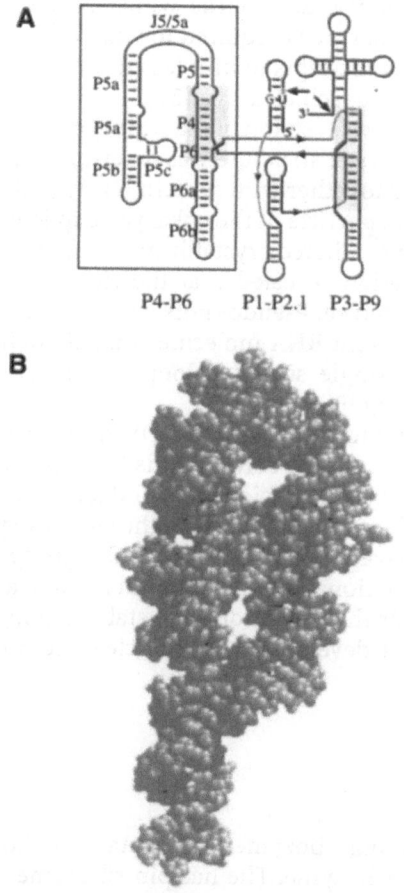

Fig. 2. The P4-P6-domain of the group I intron of *Tetrahymena thermophila*. **A** Schematic representation of the secondary structure of the whole self-cleaving intron (modified after Cate et al. [34]). The labels for the paired regions P4 to P6 are indicated. The grey shaded region indicate the phylogenetically conserved catalytic core. The portion of the ribozyme that was crystallized is framed. **B** Three dimensional structure of the P4-P6 domain. Helices of the P5abc extension are packed against helices of the conserved core due to a bend of approximately 150° at one end of the molecule

the recognition motif for the GAAA-tatraloop. The metal-binding within this ribozyme fragment was investigated as well [35, 36]. It appeared that the P4-P6-domain contained a core of five Mg^{2+}-ions and it was suggested that this might be the RNA-version of a hydrophobic core which is similar to analogous elements in proteins.

2.1.3
The Hepatitis Delta Virus Ribozyme

The hepatitis delta virus (HDV) ribozyme is part of the circular single stranded RNA genome of the hepatitis delta virus which consists of a total of 1700 nucleotides. The HDV ribozyme is required for the processing of multimers of the genomic linear RNA transcripts to unit length by catalyzing a transesterification reaction that results in self cleavage [23].

A 72-nucleotide self-cleaved version of that ribozyme was the third large catalytic RNA from which a crystal structure was determined recently (Fig. 3) [37]. The RNA was crystallized together with protein U1 A, a subunit of the U1 small nuclear ribonucleoprotein particle of the eukaryotic splicing machinery [53], to which it binds and which facilitated crystallization without affecting its activity. The structure determination revealed that the ribozyme folds into a nested double "pseudoknot" structure. Pseudoknots are widespread structural motifs in many functionally different RNA molecules that are defined by the Watson-Crick base-pairing of a single stranded loop region with a complementary sequence outside this loop [40, 54].

The double pseudoknot fold enables the HDV ribozyme to form a deep cleft protecting the active site from the solvent. This had been suggested previously by Rosenstein and Been based on biochemical data [55]. The X-ray structure established that buried deeply in this cleft lies the 5'-hydroxyl leaving group that results from the self-cleavage reaction. This 5'-OH group is surrounded by a number of important functional groups required for activation of the attacking 2'-hydroxyl group and for the transition state stabilization by neutralization of the negative charges that develop during the cleavage process in the leaving group.

2.1.4
The Hairpin Ribozyme

Another naturally occurring ribozyme which catalyzes phosphodiester transfer reactions is the hairpin ribozyme. The hairpin ribozyme has been the subject of a number of excellent review articles [24, 25]. Several independent studies performed recently have indicated that the hairpin ribozyme has an interesting feature which distinguishes it from the aforementioned ribozymes mechanistically: While the HHR, the group I intron, the HDV ribozyme and many other ribozymes that we are going to meet in this review are metalloenzymes and require divalent metal ions in their active sites for functional group activation, divalent metals ions only play a passive role (they are mainly required for cor-

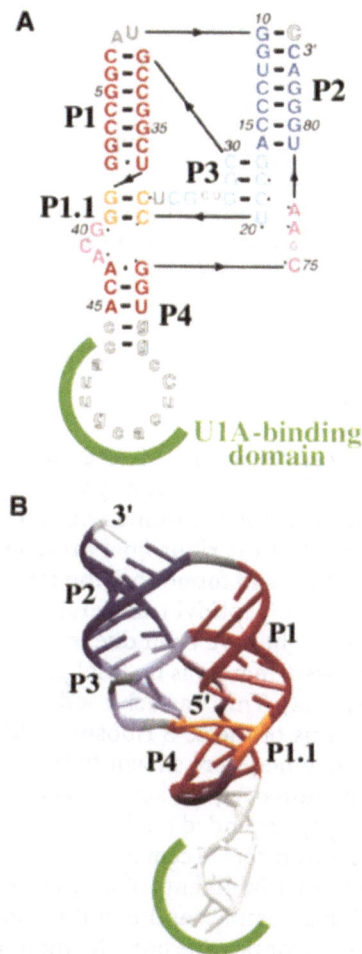

Fig. 3. The hepatitis delta virus ribozyme. A Secondary structure of the genomic HDV ribozyme RNA used for the determination of the crystal structure [37]. The color code is reflected in the three dimensional structure **B** of this ribozyme. P1 to P4 indicate the base-paired regions. Nucleotides in small letters indicate the U1 A binding site that was engineered into the ribozyme without affecting the overall tertiary structure. The yellow region indicates close contacts between the RNA and the U1 A protein

rect folding of the nucleic acids) in the cleavage reaction catalyzed by the hairpin ribozyme (Fig. 4) [56–58].

This is also true for a number of in vitro selected DNA enzymes which were selected under divalent metal-free buffer conditions [59, 60]. These results contradict the common assumption that all ribozymes are metalloenzymes and provide a number of ribozymes for which it will be very interesting to determine their exact catalytic mechanisms at high resolution.

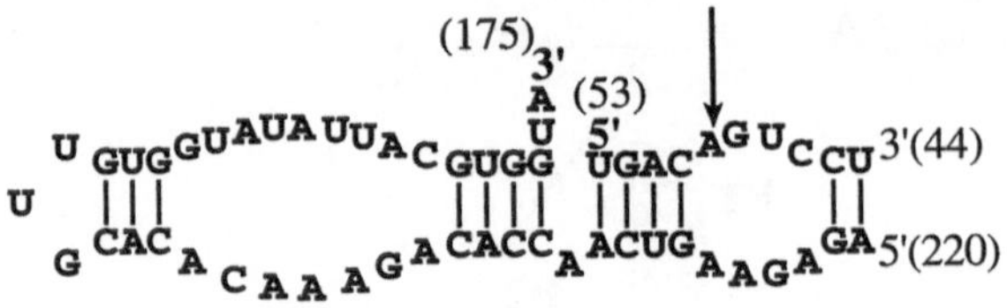

$$(175)_{3'}$$
$$\text{A } (53)$$
$$\text{U}$$
$$\text{U } \text{GUG}^{\text{GU}^{\text{AUA}}\text{UU}}\text{A}_{\text{C}}\text{GUGG} \quad ^{5'}\text{UGAC}^{\text{A}\text{G}\text{U}}\text{C}_{\text{CU}}\text{3'(44)}$$
$$\text{U } \quad | \; | \; | \quad\quad\quad | \; | \; | \; | \quad | \; | \; | \; | \quad\quad | \; |$$
$$\text{G } \text{CAC}_{\text{A}}\text{C}_{\text{A A A}}\text{G}^{\text{A}}\text{CACC}_{\text{A}}\text{ACUG}_{\text{A A G A}}^{\text{GA}}\text{5'(220)}$$

Fig. 4. The self-cleaving "hairpin" motif from the satellite RNA of tobacco ringspot virus (sTobRV). The *arrow* indicates the cleavage site. The numbers in brackets indicate the nucleotide positions within the sTobRV satellite RNA

2.2
Ribosomal RNA as a Catalyst

In 1993, Noller and his colleagues published a frequently cited study in which it was suggested that the 23 S ribosomal RNA might be capable of performing the peptidyl transferase activity of the ribosome without any ribosomal proteins [61]. They had removed most of the ribosomal proteins from the RNA portion by extensive phenol extraction and found that the remaining nucleic acid-containing aqueous phase retained peptidyl transferase activity. As it could not be excluded that this activity might have been due to some residual traces of protein that could not be removed, there was no final proof that peptidyl transfer is an RNA-catalyzed process [62]. This was achieved recently by testing six individually synthesized domains of the 23 S ribosomal RNA [63, 64]. These fragments, when complexed together, were shown to be capable of performing the peptidyl transfer reaction. Moreover, the authors were able to demonstrate that the catalytic activity is largely depended on the presence of domain V, which lies in the heart of the peptidyl transferase center of the 23 S rRNA. This reconstitution experiment suggested that fragments of an RNA molecule have the ability to associate into a functional complex and that it is indeed the RNA portion in 23 S rRNA that is responsible for peptide bond formation in the ribosome. As we will discuss below, additional support for this notion comes from a study by Zhang and Cech who recently succeeded in the in vitro selection of a ribozyme that catalyzes a reaction analogous to the 23 S ribosomal peptidyl transferase activity [65, 66].

3
In Vitro Selection of Catalytic Nucleic Acids

The discovery of catalytic RNAs has encouraged hypotheses about the origins of life on earth (see below) and has raised questions whether or not RNA is capable of catalyzing a much broader range of chemical reactions than those suggested by the naturally occurring ribozyme activities. One extremely powerful tool to answer these questions has been provided in the form of the in vitro selection technology. By in vitro selection, a number of novel ribozymes were selected which impressively enlarge the spectrum of chemical transformations catalyzed

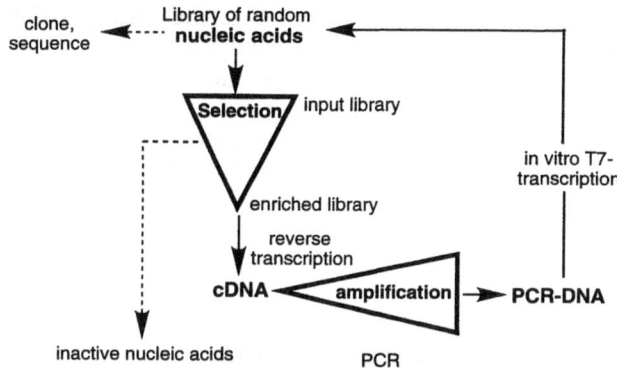

Fig. 5. Selection scheme for the in vitro selection of RNA libraries. The RNA library is subjected to a selection criterion suitable for the enrichment of functionally active sequences. The few selected individual sequences are amplified by reverse transcription (RT) and polymerase chain reaction (PCR). The PCR-DNA is then subjected to in vitro transcription with T7 RNA polymerase. The resulting enriched and amplified RNA library can be used as the input for the next selection cycle. This process is repeated until active sequences dominate the library. At this point, individual sequences can be obtained by cloning and their sequence can be determined by sequencing

by RNA or single-stranded DNA molecules. Several excellent review articles summarize them [67–75].

In vitro selection is a combinatorial approach in which functional molecules are selected from large libraries of randomized RNAs or DNAs by selection techniques that are suitable for the enrichment of a particular property such as the binding to a target molecule or a particular catalytic activity (Fig. 5).

The starting pool is generated by standard DNA-oligonucleotide synthesis by which up to 10^{15} different DNA molecules can be synthesized at once. The design of pool oligonucleotides involves a completely random base-sequence which is flanked by defined primer binding sites to allow amplification by the polymerase chain reaction (PCR). The synthetic DNA can be transcribed into RNA in vitro because it contains an appropriate promoter, usually the promoter for the RNA polymerase from the phage T7. The selections are performed on the assumption that some molecules in the original pool must have the right receptor structure to bind a substrate or have the correct folding to perform catalysis of a particular chemical reaction. These rare sequences are separated from the vast majority of non-functional molecules and are amplified by the polymerase chain reaction. Since a complete enrichment of functional molecules cannot be achieved in a single step several iterative cycles of selection and amplification are required.

3.1
In Vitro Selection of Catalytic RNA

The most recent progress in this field of catalytic RNA, modified RNA, or ssDNA-selections include modified RNAs that catalyze a Diels-Alder reaction

[6] or amide bond formation [76, 77], RNAs that catalyze the formation of peptide bonds similar to the reaction catalyzed by the ribosome [65, 66], RNA and DNA sequences with various ligase activities [18, 78–83], a ribozyme for the formation of 5′-5′-diphosphate-bonds [84, 85], ribozymes for ester transfer reactions [76, 86], a ribozyme that catalyzes the formation of a glyco-sidic bond between uracil and phosphoribosyl pyrophosphate [87], and many more.

3.1.1
Indirect Selections

In vitro selection strategies can be sub-divided into two types: direct and in-direct selections. These two types of selection experiments directed at the isola-tion of synthetic catalytic nucleic acids differ mainly by their technical concept, their design and their outcome.

The idea of indirect selections goes back to John Haldane [88], Linus Pauling [89], and Bill Jencks [90]. According to them, every compound that is capable of stabilizing a transition state of a given chemical transformation should also be able to catalyze the reaction itself. Thus, if compounds that bind a transition state analog (TSA) can be isolated they can then be tested for their ability to catalyze the corresponding reaction. This principle has been applied successful-ly for the isolation of catalytic antibodies which can be obtained by immunizing animals with transition state analogs [91]. After a population of TSA-binding antibodies has been obtained, each antibody-clone is screened for catalytic activity in subsequent experiments.

While indirect selections work quite well for antibodies they have been less successful in the case of catalytic nucleic acids. There are only three examples which prove that it is possible in principle to obtain a ribo- or deoxyribozyme by selecting an aptamer that binds to a TSA: A rotamase ribozyme [7], a ribo-zyme capable of catalyzing the metallation of a porphyrin derivative [92], and one catalytic DNA of the same function [93]. Another study reported the selec-tion of a population of RNA-aptamers which bind to a TSA for a Diels-Alder reaction but the subsequent screen for catalytic activity was negative for all indi-vidual RNAs tested [94]. The attempt to isolate a transesterase ribozyme using the indirect approach also failed [95].

Following the TSA-based strategy, RNA aptamers were selected that specifi-cally complexed the TSA for the isomerization of an asymmetrically substituted biphenyl derivative (Scheme 1) [7]. The selection was performed by affinity chromatography of a randomized pool on the TSA immobilized on agarose. After seven rounds of selection, the RNA pool accelerated the basal reaction 100-fold and was completely inhibited by the planar TSA.

The second example of a catalytic RNA obtained by the indirect selection approach is the isolation of a 35 nucleotide RNA molecule which binds meso-porphyrin IX and catalyzes the insertion of Cu^{2+} into the porphyrin with a value of k_{cat}/K_M of 2100 $M^{-1} s^{-1}$ [92]. Remarkably, the k_{cat}/K_M achieved by the RNA was close to that of the Fe^{2+}-metallation of mesoporphyrin catalyzed by the protein enzyme ferrochelatase.

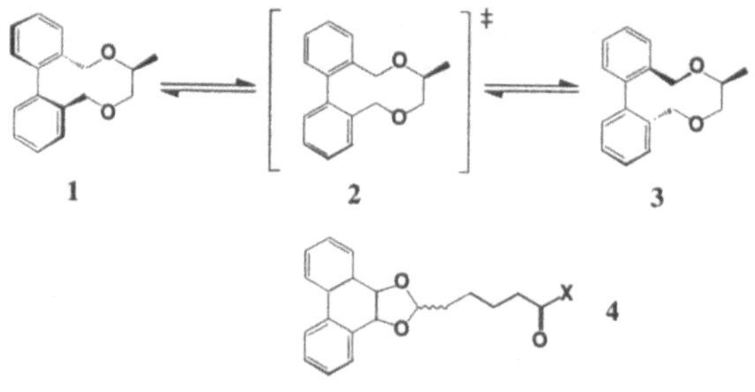

Scheme 1

The only indirect selection that led to a catalytic DNA is a deoxyribozyme that catalyzes the same class of porphyrin metallation as the aforementioned ribozyme. The ssDNA oligonucleotide showed a k_{cat} of 13 h^{-1} for the insertion of Cu^{2+} into mesoporphyrin IX [93, 96–99]. This corresponds to a rate enhancement of 1400 compared to the uncatalyzed reaction which is as good as a catalytic antibody for the same reaction.

Recently, another example of a DNA-aptamer that was selected for binding to a small molecule and that was found to accelerate weakly a chemical transformation was reported [100]. These aptamers selected to bind to the fluorophor sulforhodamine B with high affinity were also capable of promoting the oxidation of a related molecule, dihydrotetramethyl rosamine, albeit with low efficiency.

3.1.2
Direct Selections

The more successful strategy for the isolation of RNA- and DNA-based catalysts involves the direct screening of nucleic acids libraries for catalytic activity. This approach is called direct selection [6, 65, 77, 78, 86, 101–107]. In direct selections, nucleic acids that are capable of catalyzing a particular chemical transformation modify themselves with a tag or other characteristic that allows their preferential enrichment over those molecules which are catalytically inactive [108]. The design of ribozyme-selections involving reactions between two small substrates requires that one reactant be covalently attached to every individual member of the starting RNA pool. After the reaction with another substrate which usually carries the selection-tag has occurred, the self-modified RNA is immobilized on a solid support, separated from non-active molecules, and then cleaved off the support.

The principle of direct selections was first introduced by Gerald Joyce and his coworkers. From a library of mutagenized sequences of the *Tetrahymena* group I intron, a series of RNA variants were isolated which had evolved to cleave a non-natural DNA substrate corresponding to their normal RNA substrate oligonucleotides [109]. Since this pioneering work, a large number of in vitro selections with very different goals have been carried out. Thereby, not only variants of natural ribozymes with altered functionalities and substrate specificities were obtained [74, 110], but also ribo- and deoxyribozymes with completely

Table 1. Novel ribozymes and deoxyribozymes[a]

Reaction	References
Phosphor transfer reactions	
2′,5′- and 3′,5′-ligation	[78, 81]
Oligonucleotide phosphorylation	[102, 105]
Cleavage of DNA/RNA chimeric oligonucleotides	[59, 60, 111–114][c]
RNA cleavage	[104, 115], [116][c]
DNA ligation	[80][c]
5′,5′-RNA ligation	[79]
DNA cleavage	[117, 118][c]
RNA polymerization	[83, 119]
RNA "capping"	[84, 85, 120, 121]
Ligation of AMP-activated RNA	[82]
Cofactor-dependent RNA cleavage	[113, 122][c]
Other reactions	
Isomerization (Sigma rotation)	[7]
N-alkylation	[103]
Aminoacylation	[76, 86, 106, 123, 124]
S-Alkylation	[107]
Porphyrin metallation	[92, 93, 98, 99]
Amide-bond formation	[76], [77][b]
Peptidyl transfer	[65, 66]
Diels-Alder reaction	[6][b]
Nucleotide/phosphoribosyl transfer	[87]
Redox activity	[100]

[a] catalytic RNAs derived from natural ribozymes are not listed in this table.
[b] the library contained modified bases.
[c] deoxyribozyme.

novel catalytic properties [67–75]. Table 1 summarizes in vitro selected nucleic acid catalysts known to date.

3.1.2.1
Ribozymes Catalyzing Reactions at Phosphordiester Bonds

The first ribozymes evolved completely de novo was an RNA ligase [78] and a polynucleotide kinase [102, 105]. In analogy to the natural ribozymes, these RNAs catalyzed phosphodiester transfer reactions as well. For the ligase selection, a library of $>10^{15}$ different RNA sequences was screened for ribozymes which catalyze the formation of a phosphodiester bond between themselves and an external RNA oligonucleotide. The 5′-end of the library was designed as a hairpin loop motif, to be able to fold into close spatial proximity to the 3′-end of the substrate oligonucleotide. The concept was that only those sequences which catalyze the nucleophilic attack of a 2′- or 3′-hydroxy group on the adjacent 5′-triphosphate were able to transfer the substrate oligonucleotide onto themselves. Thus, active catalysts "labeled" themselves with the covalently attached oligonucleotide which enabled their separation from bulk unreacted pool RNA simply by affinity chromatography on an immobilized sequence complementary to the substrate oligonucleotide. Indeed, after ten rounds of successive selection and amplification, active ligase ribozymes became enriched. The library that now consisted mostly of active sequences catalyzed the ligation of the short RNA oligonucleotide with a rate constant of 0.06 min^{-1} which is about seven million times faster than the template directed background reaction. Cloning and sequencing revealed that several classes of ribozymes had been enriched which either catalyzed the formation of 2′,5′- or 3′,5′-phosphodiester bonds [81, 125]. Interestingly, it turned out that the selected ribozymes had evolved alternative binding sites for their substrate and did not make use of the pre-designed substrate binding site. One of the ligase ribozymes, designated as "the class I ligase", was extensively characterized and its secondary structure was elucidated [125]. Furthermore, a variant of this ribozyme was constructed which was able to perform the template directed extension of an external RNA primer by using mononucleotide triphosphates (Fig. 6) [119].

A modified version of the class I ligase which contained a different substrate hybridization site was also used for the development of a continuous RNA evolution system (Fig. 7) [83]. The hexameric substrate hybridization site of the original class I ligase, 5′-GACUGG-3′ had to be changed to 5′-UAUAGU-3′ in order to make it complementary to an oligonucleotide substrate which corresponded to the sequence of the T7-promoter. This switch in substrate specificity was necessary because the continuous evolution scheme was designed to evolve ribozymes which were capable of ligating a T7-promoter sequence onto their 5′-end with very high efficiency in the presence of reverse transcriptase, the 3′-primer, dNTPs, NTPs, and T7-RNA polymerase. In this way, a competing

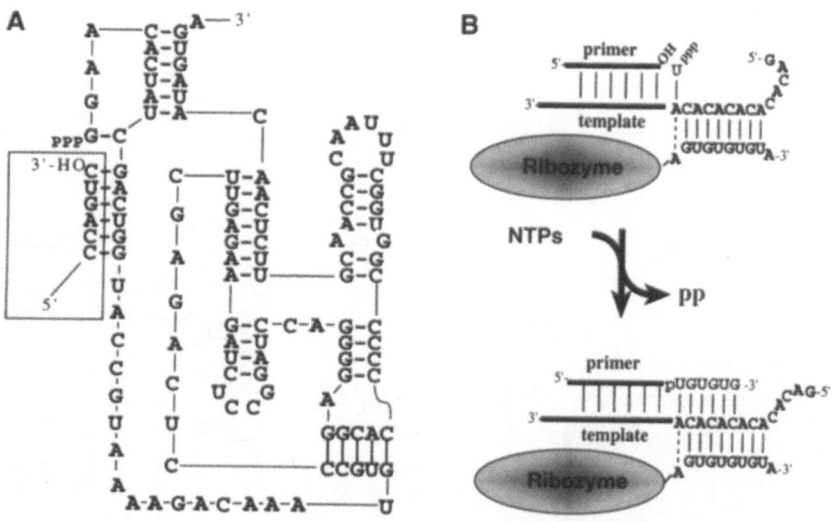

Fig. 6. A Secondary structure of the class I ligase. B Template-directed RNA polymerization of up to six nucleotides catalyzed by the class I ligase (Ribozyme)

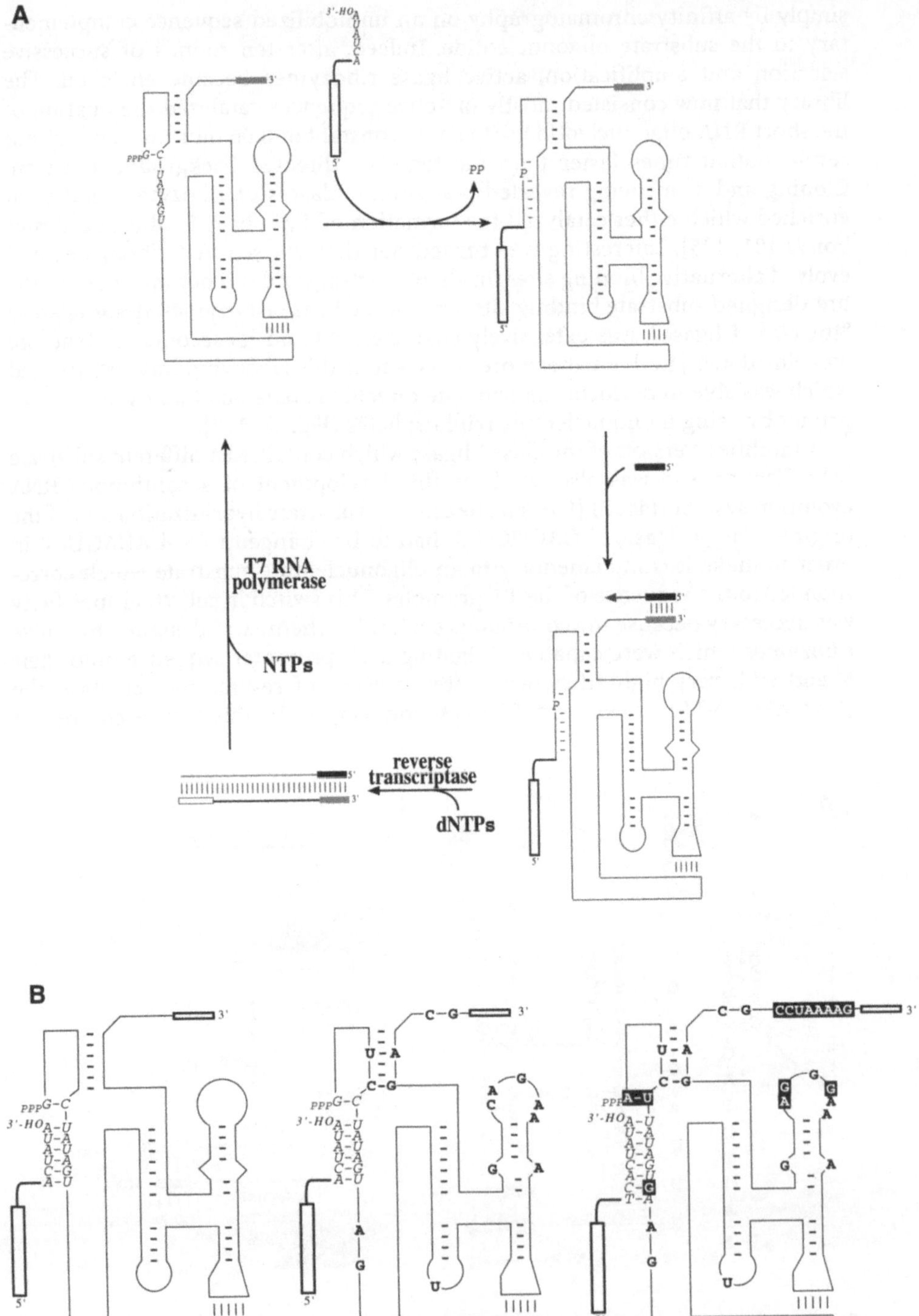

A

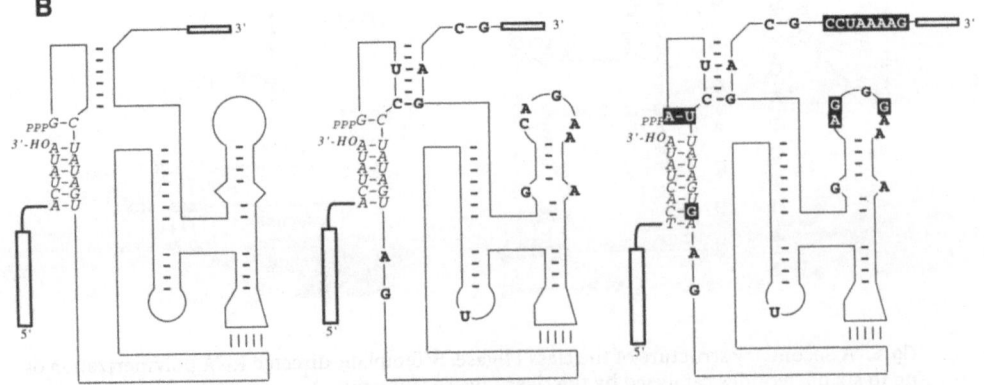

B

situation was generated, in which only those ribozymes were replicated by T7-polymerase that had ligated the T7-promoter onto their 5'-end *before* reverse transcriptase would generate too much "sense"-DNA from the ribozyme-template making catalysis impossible. With their system, Wright and Joyce have developed a continuous evolution scheme similar to the $Q\beta$-replicase system used by Spiegelman [126], with the difference that the actual catalytic step is carried out by a ribozyme. The amplification reaction, however, still depends on the "helper proteins" reverse transcriptase and T7-RNA polymerase. The continuous evolution reaction might also be used to develop a ribozyme with RNA polymerase activity, which, perhaps, some day makes the use of a proteinaceous RNA-polymerase obsolete. This process enables the replication, mutation, and selection of many ribozyme generations within a very short period of time and will be of great interest for molecular evolution studies.

In another attempt to provide a starting point for the evolution of self-replicating ribozymes, Chapman and Szostak designed a selection experiment for the generation of RNA molecules that ligate their 3'-end to a hexanucleotide with a 5'-phosphate activated as phosphorimidazolide [79] (see Fig. 8). However, the isolated ribozyme catalyzed the attack of the 5'-terminal γ-phosphate group on the 5'-phosphorimidazolide of the substrate oligonucleotide forming a 5'-5' tetraphosphate linkage. Depending on whether a 5'- mono-, di-, or triphosphate was present, the 54 nucleotide long pseudoknot motif was also capable of generating di- or triphosphate linkages.

Hager and Szostak used an RNA library in which each member was "capped" by an adenosine-5'-5'-pyrophosphate group at the 5'-end to isolate ribozymes that catalyze the ligation of an oligoribonucleotide to this activated group. This reaction results in the formation of a 3'-5'-ligation and the release of AMP [82].

A ribozyme activity that led to RNA-modifications that are analogous to the 5'-5' pyrophosphate "caps" of eukaryotic RNA transcripts was selected by Huang and Yarus [84]. Actually the author's intention was to isolate ribozymes which catalyze the formation of a mixed anhydride between an amino acid carboxylate and a 5'-terminal phosphate of an RNA, an activity that is chemically analogous to the activation of amino acids by ATP catalyzed by aminoacyl tRNA synthetases. However, while the selected ribozymes did

----

Fig. 7. Continuous evolution of the class I ligase ribozyme. A Schematic for the continuous evolution system leading to enrichment of a highly active class I ligase. B *Left panel:* schematic secondary structure of the class I ligase with variant substrate binding specificity (sequences shown) compared to the original class I ligase. Because this change in substrate specificity resulted in a 1000-fold reduced catalytic activity the new class I ligase construct had to be evolved by stepwise and rapid in vitro evolution to improve the ligation rate. From this, ribozymes emerged which were not only capable of performing fast enough catalysis with the modified substrate hybridization sequence, but also of accepting a DNA-3'-r(N_4)-substrate. A ribozyme which was mutated at 17 positions (mutations shown) compared to the starting sequence was used to start the continuous evolution reaction (*middle panel*). The *right panel* shows those positions which had changed after continuous evolution

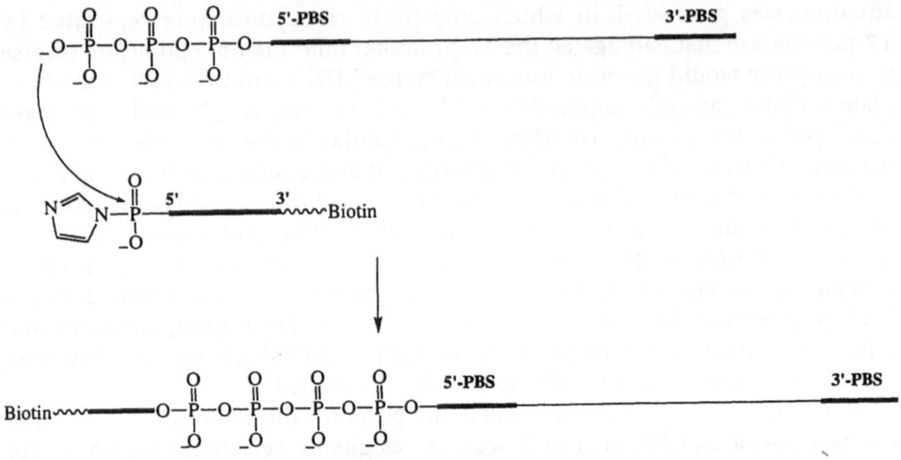

Fig. 8. Reaction catalyzed by the RNA to generate a 5′-5′-tetraphosphate linkage

catalyze a reaction that led to the generation of pyrophosphate – the expected by-product of an aminoadenylation reaction – Huang and Yarus showed in a set of control experiments that their activity was independent of the presence of amino acids. Further characterization established that the ribozymes catalyze the formation of 5′-5′-polyphosphate linkages employing substrate molecules such as oligonucleotides or nucleotide mono-, di-, or triphosphates that contain 5′-terminal phosphate groups. Even 5′-phosphate containing biological cofactors such as FMN, NADP+, CoA, PRPP, or thiamine pyrophosphate could be utilized as substrates. These ribozymes also promoted the competing hydrolysis of phosphoanhydrides [85, 120]. In a recent study it was shown that one of the ribozymes, when engineered properly can also use two small-molecule substrates and ligate them together in an intermolecular reaction [121].

3.1.2.2
Ribozymes Catalyzing Non-Phosphodiester Chemistry

Many examples of catalytic nucleic acids obtained by in vitro selection demonstrate that reactions catalyzed by ribozymes are not restricted to phosphodiester chemistry. Some of these ribozymes have activities that are highly relevant for theories of the origin of life. Hager et al. have outlined five roles for RNA to be verified experimentally to show that this transition could have occurred during evolution [127]. Four of these RNA functionalities have already been proven: Its ability to specifically complex amino acids [128–132], its ability to catalyze RNA aminoacylation [106, 123, 133], acyl-transfer reactions [76, 86], amide-bond formation [76,77], and peptidyl transfer [65,66]. The remaining reaction, amino acid activation has not been demonstrated so far.

The central role aminoacylated RNAs play in translation processes suggests that acyl transfer reactions catalyzed by RNA might have facilitated the development and optimization of the translation apparatus during early evolution. The three ribozymes described below expand the scope of RNA catalysis towards this direction.

Lohse and Szostak recently described an in vitro selected ribozyme which catalyzes the transfer of a biotinylated methionine residue from an oligonucleotide substrate to the 5′-OH of a pool RNA molecule [76]. Their design of the acyl donor molecule involved the linkage of the amino acid to the 3′-end of a short 6-mer oligonucleotide capable of hybridizing to the ribozyme. By substituting the 5′-OH group of the ribozyme for an amino group the ester transferase could be engineered to perform a corresponding transfer of the amino acid to the 5′-NH$_2$-modified ribozyme, resulting in the first example of an RNA that catalyzes amide-bond formation.

A second example of a similar activity is a ribozyme which catalyzes the transfer of an amino acid ester from a biotinyl-N-phenylalanyl-2′(3′)-adenosine-5′-monophosphate (Bio-Phe-AMP) substrate onto a specific ribose 2′-OH group (Fig. 9) [86] This ester transferase ribozyme is thus an example of an RNA which catalyzes a reaction at a carbon center by utilizing a low molecular weight cofactor. The reaction depends strongly on the presence of divalent metal ions and can be inhibited by AMP, but not with GMP, indicating that a specific binding pocket for the Bio-Phe-AMP substrate exists. The transformation reaches equilibrium due to a significant level of the corresponding reverse reaction, 2′(3′)-aminoacylation of AMP.

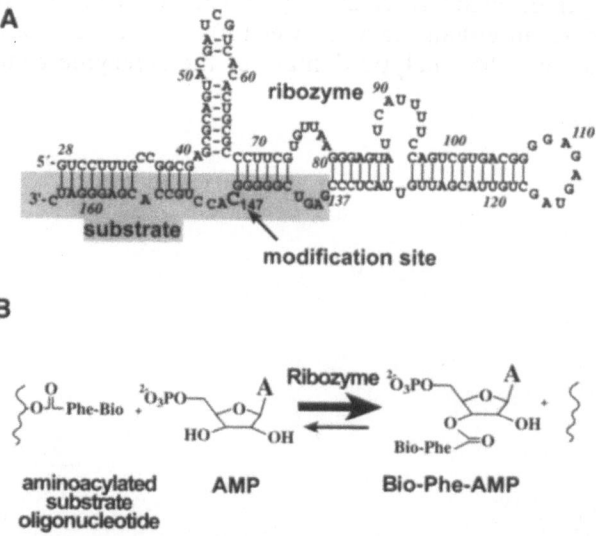

Fig. 9. The transacetylase ribozyme. A Secondary structure of the clone 11 transacylase ribozyme based on the Zuker RNA folding algorithm Mfold. The oligonucleotide substrate is shaded in *gray*. The 2′-OH group of cytosine 147 (*arrow*) is the site of modification of the oligonucleotide substrate. B Reaction catalyzed by the clone 11 transacylase ribozyme. Note that the equilibrium of the reaction lies strongly on the side of the Bio-Phe-AMP substrate

A prerequisite for peptide and protein synthesis in all modern-day life forms is the formation of aminoacyl-tRNAs catalyzed by aminoacyl-tRNA synthetases. These enzymes first activate the α-carboxy group of the amino acid by forming an aminoacyl-adenylate containing a highly activated mixed anhydride group which is then used to transfer the amino acid to the 3′(2′)-hydroxy terminus of the cognate tRNA. Illangasekare et al. used an in vitro selection strategy to obtain an RNA that catalyzes the esterification of an activated phenylalanine to its own 3′(2′)-end [106, 123]. An RNA library consisting of $\sim 10^{14}$ different molecules was incubated with synthetic phenylalanyl-5′-adenylate. RNA molecules which had catalyzed their own aminoacylation thus became self-modified with a free α-NH_2-group from the amino acid. This nucleophilic amino group was selectively reacted with the N-hydroxysuccinimide of naphthoxyacetic acid. Thus, only those RNAs with the amino acid covalently attached to themselves contained the naphthoxy residue and therefore differed significantly in their hydrophobicity from inactive molecules allowing their separation by reversed phase HPLC (Fig. 10). Eleven cycles of selection resulted in a variety of self-aminoacylating ribozymes of which the most active showed a rate enhancement of 10^5-fold compared to the background rate.

There are three examples which show that RNAs can be selected that catalyze amide bond- [76, 77] and even peptide bond formation [65, 66]. Among of them is also the first example in which a modified RNA library was used for the selection of a catalytic RNA. In order to enlarge the functional group diversity Wiegand et al. replaced all uridine residues in an RNA library for 5-imidazolyl uridines thus mimicking histidine, a building block in proteins that is capable of performing general acid base catalysis [77]. The isolated amide synthetase exhibited a catalytic rate enhancement between 10^4- and 10^5-fold for the amide bond forming reaction. The catalytic domain of the ribozyme contained three 5-

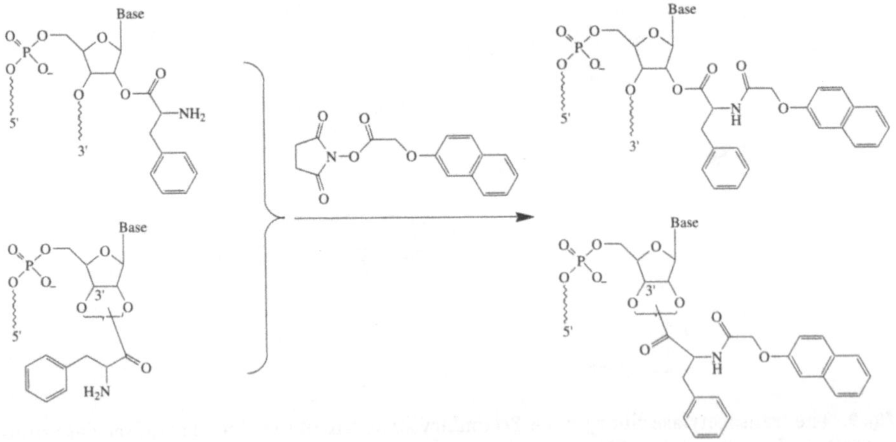

Fig. 10. 2′- and 2′,3′-aminoacylation of catalytic RNAs. Catalytic RNAs that can self-aminoacylate can be distinguished by non-modified ones by their increased hydrophobicity by the naphthoxyacetyl label

imidazolyl uridine residues which were absolutely necessary for activity as replacement of these modified residues with uridines abolished catalytic activity. This study also showed that the lack of functional group diversity in nucleic acids can easily be overcome chemically by applying chemically synthesized nucleotide triphosphate (NTP) derivatives that contain additional functionalities [134]. The only requirements with respect to the selection process are that the chemically modified NTPs are accepted as substrates by the nucleic acid replicating enzymes and that the modification does not interfere with the ability of the NTP to pair in the Watson-Crick sense. Researchers have investigated the ability of natural polymerases to accept a number of synthetic dNTPs as substrates [135]. For example, in a recent study Barbas and Sakthivel introduced various functional groups that are normally not found in a nucleic acid, such as carboxylic acids, imidazoles, amines, phenols, and pyridines which were introduced into uridine triphosphates at the 5-position of the base via a rigid spacer arm [136].

Zhang and Cech used a library of 1.3×10^{15} different RNAs with a length of 196 nucleotides to isolate novel peptidyl transferase ribozymes [65, 66]. For their direct selection scheme they coupled the α-carboxy group of a phenylalanine covalently to the 5'-end of all members of the RNA library. This modified pool was then incubated with a synthetic molecule that mimicked the 3'-end of a tRNA aminoacylated with the formyl methionyl residue. In their minimal version of a formylmethionyl tRNA, the formyl group was substituted by a biotin group and the tRNA was minimized to a single 3'-adenosine residue. After 19 cycles, ribozymes were obtained which catalyzed the formation of a peptide bond between the methionin carboxy group and the amino group of phenylalanine – in analogy to the peptidyl transfer mechanism in the ribosome (Fig. 11). Some of the selected ribozyme variants observed rate constants (k_{obs}) of almost 0.1 min^{-1}, which translates into a rate enhancement of 10^5-fold compared to the uncatalyzed reaction. The amino acid/AMP ester is mainly bound to the ribozyme by the adenosine moiety independently of which amino acid is used, allowing amino acids other than methionine to be utilized in this reaction [66].

Eaton and his colleagues have demonstrated that it is also possible to obtain ribozymes that catalyze the formation of carbon-carbon bonds by isolating a "Diels-Alderase"-ribozyme. While the indirect selection approach failed [94] the ribozyme isolated in a direct selection accelerates the Diels-Alder reaction between a diene which is covalently attached to its 5'-end via a long polyethylene glycol linker and a biotinylated maleimide dienophile by a factor of 300 [6]. As with their previous example of amide-bond forming ribozymes, a modified RNA library was used in which all uracil residues were substituted by pyridyl methyl-modified UMPs. Interestingly, the selected Diels-Alderase ribozyme is also dependent on the presence of Cu^{2+}, as was the amide synthetase. Cu^{2+} may either be required for forming the proper structure of the catalyst and/or may actively participate in the catalytic step by providing Lewis acid sites which are known to be advantageous for the acceleration of Diels-Alder reactions in water [137].

One of the most interesting ribozymes isolated by in vitro selection techniques is a novel catalytic RNA that promoted the formation of a glycosidic bond

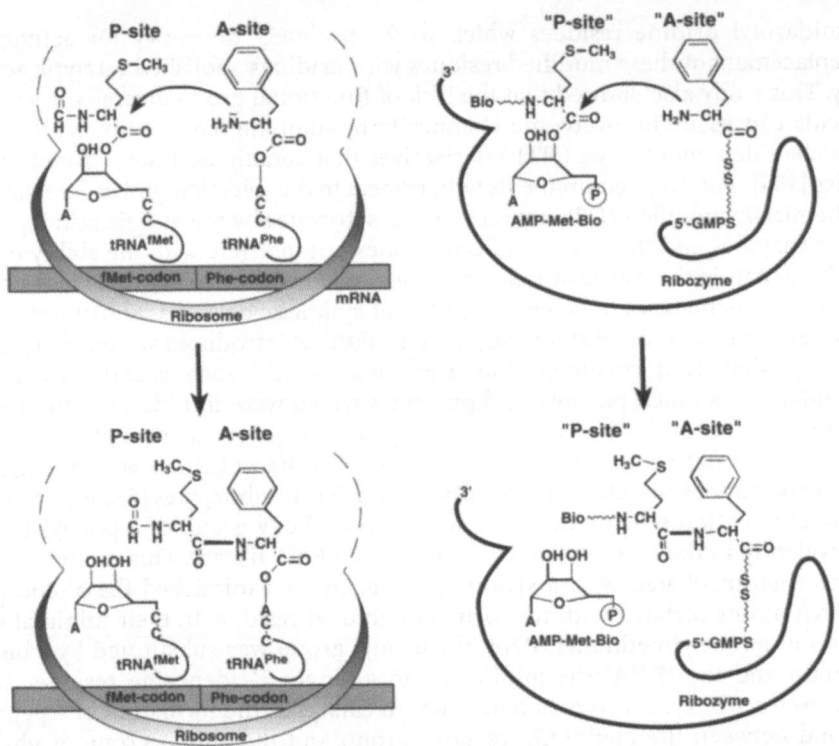

Fig. 11. Comparison of the peptidyl transfer reaction in the ribosome and in the selected peptidyltransferase ribozyme. The ribosome contains a binding site for the peptidyl-tRNA (P-site) and for the aminoacyl-tRNA (A-site). In the selected ribozyme the binding site for the AMP-Met-Bio substrate would be analogous to the P-site. The attacking α-amino group which is bound in the A-site in the ribosome is covalently attached to the 5'-end in the ribozyme. Catalytically active RNAs preferentially attach the biotin tag onto themselves and can thus be separated from the inactive ones

between 4-thiouracil (4-S-Ura) and 5'-phosphoribosyl-1'-pyrophosphate (PRPP), linked to the 3'-end of an RNA library [87] (Scheme 2).

While the PRPP substrate was covalently attached to the RNA pool, the 4-S-Ura substrate was incubated with the library in solution. Active molecules were selected with a special gel electrophoretic technique on the basis of the

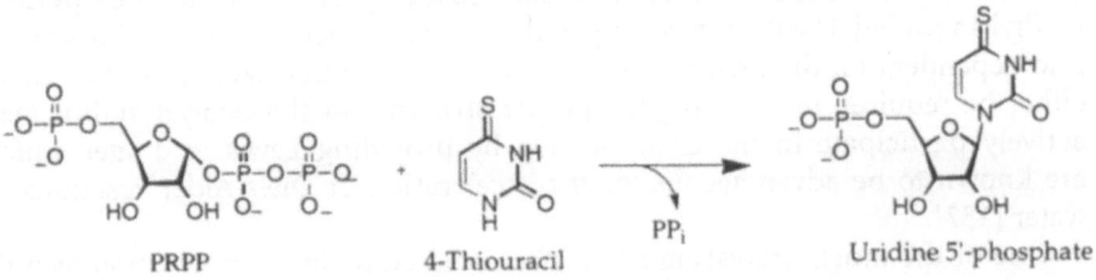

PRPP   4-Thiouracil   Uridine 5'-phosphate

Scheme 2

attached thio group [138]. Nucleotide formation catalyzed by the selected catalytic RNAs was at least 10^7 times faster than the uncatalyzed reaction. The ribozymes showed remarkable specificity for the 4-S-Ura substrate. No reaction occurred with 2-thiouracil, 2,4-thiouracil, 2-thiocytosine, 2-thiopyrimidine, 2-thiopyridine or 5-carboxy-2-thiouracil. This study showed that RNA can perform reactions by binding substrates that are smaller than purine nucleotides. The study by Unrau and Bartel is therefore an important contribution to the RNA-world hypothesis which requires that ribozymes would have needed to promote many metabolic reactions which involve organic molecules of low molecular weight.

3.1.3
Direct Selections From Aptamer-Based Libraries

Some of the ribozyme selections discussed above require the reactants to be attached covalently to the nucleic acid. In other cases, substrates were incubated with the library in solution and the resulting ribozymes had to evolve defined binding sites for the substrates and to provide the scaffold for its correct positioning. The isolation of new ribozymes might be facilitated by first selecting an aptamer-sequence for binding to a cofactor needed in the reaction and including a highly mutagenized aptamer sequence in addition to a completely randomized portion as a basis for the selection of functional molecules. This method of direct selection may provide a more focused sequence space to increase the number of sequences in the library that are capable of binding to the substrate.

Lorsch and Szostak [102] generated an RNA-library in which the central sequence of a partly randomized ATP-binding RNA aptamer, which had previously been selected by affinity chromatography on ATP agarose [139, 140] was surrounded by three completely randomized regions of a total of 100 bases. This library was used for the selection of an ATP-dependent oligonucleotide kinase by incubation with ATP-γ-S [102, 105]. RNA molecules onto which the γ-thiophosphate group of the ATP-γ-S had been transferred could be separated from the rest of the library on activated thiopropyl agarose, since they specifically formed a disulfide bond between the thiophosphate group and the agarose. These covalently bound RNAs were then eluted by washing with an excess of 2-mercaptoethanol, amplified and reselected. After thirteen cycles of selection, seven classes of ribozymes were characterized which catalyzed various reactions. Five of these RNA classes catalyzed the transfer of the γ-thiophosphate onto their own 5′-hydroxyl group and are therefore 5′-kinases. The other two classes phosphorylated the 2′-hydroxyl groups of specific internal base positions. Starting from one of these sequences, a ribozyme was developed which can phosphorylate the 5′-end of oligonucleotides in an intermolecular reaction.

An analogous strategy was used by Wilson and Szostak to isolate self-alkylating ribozymes using an iodoacetyl derivative of the cofactor biotin [103]. After isolating biotin-binding RNA aptamers by repeated rounds of affinity chromatography and amplification, a second library was generated which consisted of the mutagenized aptamer sequence flanked on either side by 20 random nucleotides. Molecules from this library which were able to self-alkylate with the

biotin derivative were separated from inactive sequences by binding to streptavidin. By the seventh round of selection, more than 50% of the RNA performed the self-biotinylation reaction. The sequencing of individual clones
revealed that a majority of the ribozymes was derived from a single ancestral
sequence. To optimize the activity, a third selection was carried out in which the
incubation time as well as the concentration of the cofactor were progressively
lowered. The resulting ribozyme alkylated itself at the N7 position of a specific
guanosine residue within a conserved region with a rate acceleration of 2×10^7
compared to the uncatalyzed reaction.

The self-biotinylating ribozymes which originated from a biotin-binding
aptamer show an astounding structural change compared to their ancestor. A
highly conserved nucleotide stretch of the biotin binder which seems to directly mediate the interaction between biotin and the aptamers as well as the catalytically active molecules is retained in the self alkylating ribozymes with only a
single point mutation. Yet, the secondary structural context of this consensus
sequence is highly unrelated in the two classes of molecules. While the biotin
aptamer contains a pseudoknot motif, the ribozyme forms a cloverleaf which
resembles the structure of tRNAs.

The fact that, in both selection experiments, new solutions regarding the
structure of the functional molecules have been adopted demonstrates that
the best sequence for binding is not necessarily the best sequence for performing catalysis. It seems likely that many of the sequence solutions could
also have been selected from completely randomized pools. This notion is confirmed by the aforementioned study by Hager and Szostak [82], in which the
mutated ATP-aptamer motif was also included in the starting library but
where the resulting ribozyme had no relationship to the parent ATP-binding
motif.

3.2
In Vitro Selection of Catalytic DNA

A new dimension in the development of nucleic acid based catalysts was introduced by Breaker and Joyce in 1994 when they isolated the first deoxyribozyme
[111]. It is not unexpected that DNA is also able to catalyze chemical reactions
because it was shown previously that ssDNA aptamers which bind to a variety of
ligands can be isolated by in vitro selection [141]. In the meantime, several
deoxyribozymes have been described which expand the range of chemical
transformations accelerated by nucleic acid catalysts even further and raising
question whether even catalytic DNA might have played some role in the prebiotic evolution of life on earth [69–71].

Breaker and Joyce selected a deoxyribozyme that specifically cleaved the
phosphodiester bond of a single ribonucleotide embedded within an all DNA
oligonucleotide [111]. The library of single stranded DNA contained the single
ribonucleotide at a specific position within a primer binding site to avoid the
possible effect of RNA on catalysis. Guided by the assumption that a DNA-
dependent cleavage at neutral pH would require a metal ion as a cofactor, they
included Pb^{2+} in their selection. The starting pool contained about 10^{14} double-

stranded DNAs with a randomized region of 50 nucleotides. One of the strands had a biotin group covalently attached to its 5′-end, followed by a 43 nucleotide position of defined sequence which also contained the cleavage site in the form of a single ribose adenosine unit. The biotin moiety was used to immobilize the single-stranded DNA on streptavidine agarose which was then incubated in Pb^{2+}-containing buffer. In a small fraction of active sequences, this treatment resulted in the cleavage of the phosphodiester bond at the ribose phosphate "point of fracture". Because these sequences were separated from the biotin anchor group after cleavage, they could be partitioned from the inactive ones by washing the streptavidine matrix, Fig. 12.

After five cycles of selection and amplification, a population of single-stranded DNAs was enriched that catalyzed the Pb^{2+}-dependent cleavage at the ribose residue. This intramolecular cleavage activity was transformed into an intermolecular reaction by separating the 38-nucleotide long catalytic domain from the 21-mer substrate which was cleaved specifically and with high turnover rates. Remarkably, the deoxyribozyme can perform well only with the special DNA/RNA chimeric oligonucleotide substrate and cannot cleave a pure RNA substrate of the same sequence.

This goal was recently achieved by Santoro and Joyce who isolated a deoxyribozyme which is able to hybridize to and cleave any RNA sequence under phy-

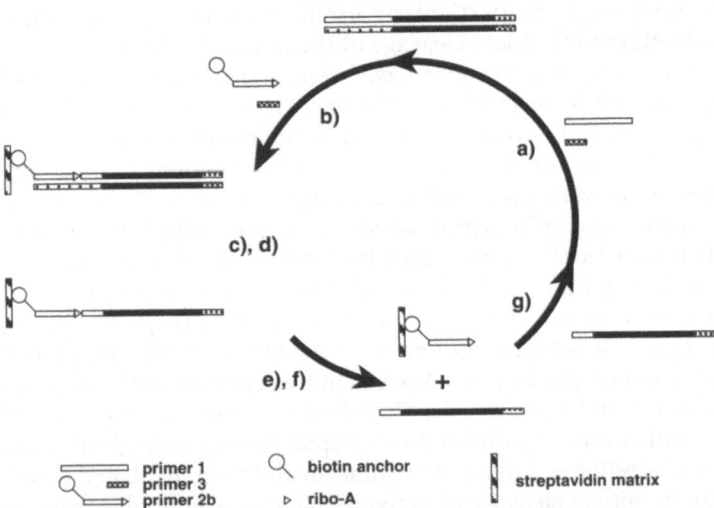

Fig. 12. Selection scheme for deoxyribozymes which cleave RNA/DNA chimeric oligonucleotides. In a first PCR a), the starting pool was amplified using primer 1 and primer 3. b) In a second PCR, the 5′-end of the pool was biotinylated and the ribonucleotide serving as cleavage site was introduced using primer 2b and primer 3. The double-stranded pool was then loaded on a streptavidin column c). The antisense strand is removed by raising the pH of the solution d) and the remaining pool of single-stranded DNA allowed to fold e) by rinsing the column with equilibration buffer. The equilibration was followed by addition of cleavage buffer f). Cleavage products were eluted from the column g) and amplified by PCR using the same primers as above

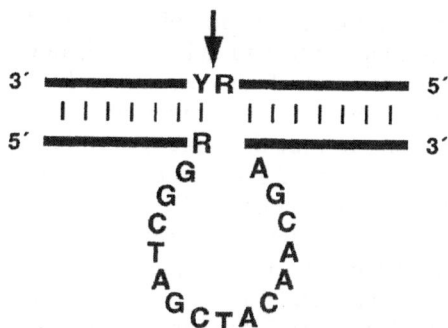

Fig. 13. The deoxyribozyme (*bottom strand*) hybridizes to the RNA substrate oligonucleotide (*top strand*). The site of cleavage is indicated by the *arrow* (R = A or G; Y = U or C). The sequences of the helical parts of the enzyme can be chosen as desired, so that almost any RNA sequence can be targeted by the catalytic DNA

siological conditions with remarkable efficiency [116]. The enzyme consists of a 15-mer catalytic domain which is flanked by sequences that can hybridize to the RNA target through Watson-Crick pairing (Fig. 13). Because of its small size and its high cleavage efficiency under multiple turnover conditions, it is a very attractive candidate for therapeutic and biotechnological applications. This catalytic DNA might be particularly useful when incorporated into antisense oligonucleotides to facilitate cleavage of the targeted mRNA.

Meanwhile other experiments have been carried out to isolate DNA molecules with either RNA- or DNA-phosphoesterase activity [69–71, 142, 143]. Since Mg^{2+}-dependent rather than Pb^{2+}-dependent cleavage is compatible with intracellular conditions and thus, more suitable for possible medical applications, deoxyribozymes were selected that used Mg^{2+} instead of Pb^{2+} for cleavage [112]. One optimized deoxyribozyme that emerged in this selection showed a cleavage rate of 0.01 min^{-1} and was also capable of intermolecular cleavage.

The catalytic potential of DNA was further examined by Carmi et al. with the same in vitro selection strategy to generate catalytic DNAs that facilitate DNA-cleavage by a redox-dependent mechanism [117, 118]. The design of this selection was based on the fact that DNA is more sensitive to cleavage by oxidative mechanisms than by hydrolysis. Therefore, a library of single-stranded DNAs (ssDNA) which was immobilized on streptavidin by a covalently attached biotin-tag was equilibrated in a buffer solution and then incubated with $CuCl_2$ and ascorbate to initiate cleavage of active sequences. The pool isolated after seven rounds of selection consisted of two distinct classes of self-cleaving ssDNA molecules. While one class performed strand scission in the presence of both Cu^{2+} and ascorbate, the other class only required Cu^{2+} as a cofactor. An optimized version of one of the deoxyribozymes shows a rate enhancement of more than 10^6-fold compared to the uncatalyzed reaction.

To develop deoxyribozymes that make use of a non-metal cofactor rather than divalent metal ions for the cleavage of a ribonucleotide residue we performed an in vitro selection under conditions of low magnesium concentration, or

even without any divalent metal ions, by incubating the immobilized single-stranded DNA library in an excess of the amino acid histidine [114]. Surprisingly, non of the resulting eight classes of deoxyribozymes utilized histidine as a cofactor for cleavage. They either depended on the presence of divalent metal ions or accelerated phosphodiester cleavage even in the absence of divalent metal ions [59]. Remarkably, one of the catalysts, showed higher cleavage activity in the presence of Ca^{2+} than of Mg^{2+}, even though calcium was never present during the selection process. We hypothesized that, in this special case, Ca^{2+} might be more suitably positioned at the cleavage site than the magnesium ion. This suggestion is supported by the observation that two Ca^{2+}-ions are bound cooperatively by the catalytic DNA.

In vitro selection technology also allowed the isolation of ribozymes that utilize non-metal ion cofactors to facilitate phosphodiester cleavage. It certainly was known long before these studies that ribozymes can do that: The group I intron from *Tetrahymena thermophila* uses guanosine derivatives to initiate the first step of phosphodiester cleavage [19], i.e. the attack of the 2'(3')-OH-group of a non-covalently bound guanosine cofactor for which the ribozyme contained a specific binding pocket [144]. More recently, still using the successful selection strategy of biotin/streptavidin-immobilization of ssDNA libraries and subsequent elution of cleaved product, deoxyribozymes that cleave phosphoramidate bonds have been reported [122]. These DNA enzymes depend on the presence of trinucleotides as cofactors for cleavage and represent the first examples of DNA enzymes that need a non-metal cofactor for activity.

Another class of cofactor-dependent deoxyribozymes that use the amino acid histidine to promote phosphodiester cleavage was isolated recently by Roth and Breaker [113]. One of the isolated DNA requires L-histidine or a closely related analog to support RNA phosphodiester cleavage with a rate enhancement of nearly a million-fold over the basal substrate cleavage rate (Fig. 14).

Presumably, the L-histidine cofactor is loosely bound to the DNA-enzyme which positions its imidazole group to serve as a general base catalyst. The histidine/DNA complex thus works by a mechanism that is similar to the first

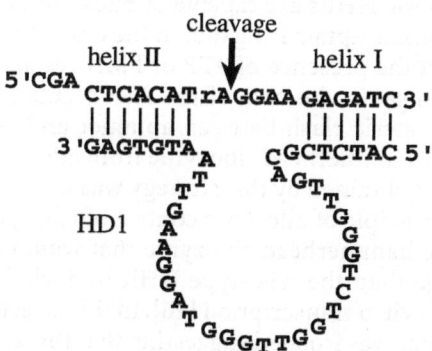

Fig. 14. Sequences and secondary structures of one of the selected L-histidine-dependent catalytic DNAs

step of the proposed catalytic mechanism of RNase A. This protein enzyme also has a histidine residue positioned in its active site which serves as a general base catalyst for RNA hydrolysis. The study by Burmeister et al. [122] and particularly the one by Roth and Breaker [113] show that catalytic nucleic acids can recruit the functional groups of complexed small organic cofactors for catalysis thereby dramatically increasing their own limited structural diversity and their catalytic potential [145].

However, even without small-molecule cofactors the catalytic scope of DNA is not restricted to phosphoresterase activity. Cuenoud and Szostak designed a selection scheme to isolate DNA molecules that catalyze the ligation of their free 5'-hydroxyl group to the 3'-phosphate group of a substrate oligonucleotide activated by imidazolide [80]. The resulting deoxyribozyme depends on the presence of a Zn^{2+} or Cu^{2+}-metal ion cofactor. The enzyme contains two conserved domains which position the 5'-hydroxyl group and the 3'-phosphorimidazolide of the substrate oligonucleotide in close proximity. Based on the selected sequences, a truncated version of the DNA ligase was designed that is able to ligate two DNA substrates in a multiple turnover reaction with a 3400-fold rate enhancement compared to the template directed ligation.

4
Allosteric Ribozymes and "Aptazymes"

One of the remarkable characteristics of ligand complexation by in vitro selected RNA aptamers is that ligand binding by the RNA molecule is always accompanied by significant structural changes in the binding RNA. Ligands seem to become an integral part of the RNA-aptamer structure once they are bound [146, 147]. This property of aptamers might have inspired the idea of fusing aptamer sequences with known catalytic RNAs to introduce the principle of allosteric regulation into ribozyme catalysis. None of the ribozymes described so far were known to operate as allosteric enzymes in vitro or in vivo.

The hammerhead ribozyme was transformed into an allosteric ribozyme by attaching the ATP- [15] or FMN-binding aptamer sequences [148–150] to its 5'-end [17]. These allosteric HHRs are capable of phosphodiester cleavage only in the absence of the cognate aptamer ligand. In the case of the conjoined aptamer/ribozyme construct the presence of ATP or FMN results in a ligand-induced conformational change. In the case of the ATP-regulated HHR-construct, ligand binding causes a steric clash between aptamer and ribozyme domains [151] which prevents the hammerhead ribozyme from adopting its active structure. The inhibition ratio obtained by this strategy was up to 180-fold. Tang and Breaker also used the principle of allosteric control of ribozyme catalysis to try to select variants of the hammerhead ribozyme that would be more active in phosphodiester cleavage than the wild-type HHR by including the inhibitory ligand ATP during the in vitro transcription [16]. In this selection, however, only the natural HHR sequence was isolated suggesting that this sequence represents a motif which has been optimized by nature for the purpose of phosphodiester cleavage.

Robertson and Ellington recently described the in vitro selection of a novel ligase ribozyme that requires allosteric activation of an oligonucleotide effector molecule for activity [18]. The allosteric ribozyme ligase is activated up to 10000-fold by the oligonucleotide effectors. "Rational" engineering of the ribozyme by incorporation of the aptameric ATP binding sequence [139] allowed its transformation into an enzyme that requires ATP in addition to the oligonucleotide effector for activity. This approach of effector-controlled ribozyme activity, when linked to RT-PCR technology, has the potential to be applied wherever accurate diagnostic quantification of oligonucleotide- or small-molecule-analytes is desired. Such allosteric ribozymes or 'aptazymes' might possess a significant application potential in medicine, diagnostics or biotechnology.

5
Ribozymes and the Origin of Life

One of the most fascinating problems many scientists, philosophers, and theologists deal with is that of the origin of life on Earth and maybe on other planets in the universe. According to the now widely accepted "RNA-world hypothesis", the way from the earliest evolving prebiotic entities to the first cells might have been accompanied by an era in which chemical transformations of primitive metabolisms were catalyzed explicitly by RNA molecules [68, 127, 152]. Most likely, life on earth did not come into existence in a single event, but emerged as the result of permanently ongoing improvement of molecules that are thought to have had the property to catalyze a wide variety of chemistries including their own synthesis. Therefore, besides their ability to self-replicate the ability to evolve in cumulative selection processes is a necessary condition for the generation of living entities from those first replicators and catalysts. The reason why RNA molecules are believed to have played a dominant role in the origin of life is that they can store information and mutate while at the same time being catalytically active. As shown by many examples of novel in vitro selected ribozymes, RNAs also might have had the potential to carry out a primitive metabolism by folding into complex three-dimensional structures that allow selective binding of other molecules and the catalysis of chemical reactions. The features of those molecules involved in the first metabolic processes were altered by mechanisms of Darwinian evolution, mutation and selection within geological timescales, as biological evolution is a slow process. In vitro selection enables the researcher to investigate the scope of RNA catalysis, or nucleic acid catalysis in general, many of which we have met in this review. As we have seen, among the artificial ribozymes that have been selected there are a number which catalyze some "key reactions of life" that have been suggested to be possible evolutionary precursors of ribosomal RNA, or RNA-based metabolism. These results provide an experimental basis for the plausibility of the RNA-world hypothesis.

6
References

1. Kruger K, Grabowski PJ, Zaug AJ, Sands J, Gottschling DE, Cech TR (1982) Cell 31:147–157
2. Gurrier-Takada C, Gardiner K, Marsh T, Pace N, Altman S (1983) Cell 35:849–857
3. Schimmel P, Alexander R (1998) Science 28:658–659
4. Piccirilli JA, McConnell TS, Zaug AJ, Noller HF, Cech TR (1992) Science 256:1420–1424
5. Dai X, De Mesmaeker A, Joyce GF (1995) Science 267:237–240
6. Tarasow TM, Tarasow SL, Eaton BE (1997) Nature 389:54–57
7. Prudent JR, Uno T, Schultz PG (1994) Science 264:1924–1927
8. Birikh KR, Heaton PA, Eckstein F (1997) Eur J Biochem 245:1–16
9. Good PD, Krikos AJ, Li SX, Bertrand E, Lee NS, Giver L, Ellington A, Zaia JA, Rossi JJ, Engelke DR (1997) Gene Ther 4:45–54
10. Christoffersen RE (1997) Nature Biotechnol 15:483–484
11. Burke JM (1997) Nature Biotechnology 15:414–415
12. Bramlage B, Luzi E, Eckstein F (1998) Trends Biotechnol 16:434–438
13. Lan N, Howrey RP, Lee S-W, Smith CA, Sullenger BA (1998) Science 280:1593–1596
14. Weatherall DJ (1998) Curr Biol 8: R696-R698
15. Tang J, Breaker RR (1997) Chem Biol 4:453–459
16. Tang J, Breaker RR (1997) RNA 3:914–925
17. Araki M, Okuno Y, Hara Y, Sugiura Y (1998) Nucleic Acids Res 26:3379–3384
18. Robertson MP, Ellington AD (1999) Nature Biotechnol 17:62–66
19. Cech TR, Zaug AJ, Grabowski PJ (1981) Cell 27:487–496
20. Jacquier A (1996) Biochimie 78:474–87
21. Pyle AM (1996) Catalytic reaction mechanisms and structural features of group II intron ribozymes, p 75–107. In: Eckstein F, Lilley DMJ (ed) Catalytic RNA, vol 10 Springer, Berlin Heidelberg New York
22. Thomson JB, Tuschl T, Eckstein F (1996). The Hammerhead ribozyme, p 173–196. In: Eckstein F, Lilley DMJ (ed) Catalytic RNA, vol 10. Springer, Berlin Heidelberg New York
23. Been MD (1994) Trends Biochem Sci 19:251–256
24. Burke JM (1994) The hairpin ribozyme, p 105–118. In: Eckstein F, Lilley DMJ (ed), Nucleic Acids Mol Biol, vol 8
25. Burke JM, Butcher SE, Sargueil B (1996). Structural analysis and modifications of the hairpin ribozyme, p 129–144. In: Eckstein F, Lilley DMJ (ed) Catalytic RNA, vol. 10. Springer, Berlin Heidelberg New York
26. Nolan JM, Pace NR (1996) Structural analysis of bacterial ribonuclease P RNA, p 109–128. In: Eckstein F, Lilley DMJ (ed) Catalytic RNA, vol. 10. Springer, Berlin Heidelberg New York
27. Saville BJ, Collins RA (1990) Cell 61:685–696
28. Eckstein F, Lilley DMJ (ed.) (1996) Catalytic RNA, vol 10 Springer Verlag Berlin
29. Uhlenbeck OC, Pardi A, Feigon J (1997) Cell 90:833–840
30. Kim S-H, Quigley GJ, Suddath FL, McPherson A, Sneden D, Kim JJ, Weinzierl J, Rich A (1973) Science 179:285–288
31. Pley HW, Flaherty KM, McKay DB (1994) Nature 372:68–74
32. Scott WG, Finch JT, Klug A (1995) Cell 81:991–1002
33. Scott WG, Murray JB, Arnold JRP, Stoddard BL, Klug A (1996) Science 274:2065–2069
34. Cate JH, Gooding AR, Podell E, Zhou K, Golden BL, Kundrot CE, Cech TR, Doudna JA (1996) Science 273:1678–1685
35. Cate JH, Hanna RL, Doudna JA (1997) Nature Struct Biol 4:553–558
36. Cate JH, Doudna JA (1996) Structure 4:1221–1229
37. Ferre-D'Amare AR, Zhou K, Doudna JA (1998) Nature 395:567–574
38. Brion P, Westhof E (1997) Annu Rev Biophys Biomol Struct 26:113–137
39. Michel F, Westhof E (1996) Science 273:1676–1677

40. Westhof E, Masquida B, Jaeger L (1996) Folding & Design 1:R78-R88
41. Tuschl T, Gohlke C, Jovin T, Westhof E, Eckstein F (1994) Science 266:785-789
42. Sczakiel G (1995) Angew Chem 107:701-704
43. Scott WG, Klug A (1996) Trends Biochem Sci 21:220-224
44. Chartrand P, Leclerc F, Cedergren R (1997) RNA 3:692-696
45. Hermann T, Auffinger P, Scott WG, Westhof E (1997) Nucleic Acids Res 25:3421-3427
46. Zhou d-M, Usman N, Wincott FE, Matulic-adamic J, Orita M, Zhang L, Komiyama M, Kumar PKR, Taira K (1996) J Am Chem Soc 118:5862-5866
47. Jaeger L, Michel F, Westhof E (1996) The structure of group I ribozymes, p 33-51. In: Eckstein F, Lilley DMJ (ed), Nucleic Acids and Molecular Biology, vol 10. Springer, Berlin Heidelberg New York
48. Abramovitz DL, Pyle AM (1997) J Mol Biol 266:493-506
49. Costa M, Deme E, Jacquier A, Michel F (1997) J Mol Biol 267:520-536
50. Jestin JL, Deme E, Jacquier A (1997) EMBO J 16:2945-2954
51. Tinoco IJ (1996) Curr Biol 6:1374-1376
52. Cate JH, Gooding AR, Podell E, Zhou K, Golden BL, Szewczak AA, Kundrot CE, Cech TR, Doudna JA (1996) Science 273:16961699
53. Sharp PA (1987) Science 235:766-771
54. Pleij CWA (1990) Trends Biochem Sci 15:143-147
55. Rosenstein SP, Been MD (1996) Biochemistry 35:11403-11413
56. Hampel A, Cowan JA (1997) Chem Biol 4:513-517
57. Nesbitt S, Hegg L, Fedor MJ (1997) Chem Biol 4:619-630
58. Young KJ, Gill F, Grasby JA (1997) Nucleic Acids Res 25:3760-3766
59. Faulhammer D, Famulok M (1997) J Mol Biol 274:188-201
60. Geyer CR, Sen D (1997) Chem Biol 4:579-594
61. Noller HF, Hoffarth V, Zimniak L (1992) Science 256:1420-1424
62. Noller HF (1993) J Bacteriol 175:5297-5300
63. Nitta I, Ueda T, Watanabe K (1998) RNA 4:257-267
64. Nitta I, Kamada Y, Noda H, Ueda T, Watanabe K (1998) Science 281:666-669
65. Zhang B, Cech TR (1997) Nature 390:96-100
66. Zhang B, Cech TR (1998) Chem Biol 5:539-553
67. Gold L, Polisky B, Uhlenbeck O, Yarus M (1995) Annu Rev Biochem 64:763-797
68. Joyce GF (1996) Curr Biol 6:965-967
69. Breaker RR (1997) Chem Rev 97:371-390
70. Breaker RR (1997) Nature Biotechnol 15:427-431
71. Breaker RR (1997) Curr Opin Chem Biol 1:26-31
72. Jaeger L (1997) Curr Opin Struct Biol 7:324-335
73. Narlikar GJ, Herschlag D (1997) Annu Rev Biochem 66:19-59
74. Pan T (1997) Curr Opin Chem Biol 1:17-25
75. Famulok M, Jenne A (1998) Curr Opin Chem Biol 2:320-327
76. Lohse PA, Szostak JW (1996) Nature 381:442-446
77. Wiegand TW, Janssen RC, Eaton BE (1997) Chem Biol 4:675-683
78. Bartel DP, Szostak JW (1993) Science 261:1411-1418
79. Chapman KB, Szostak JW (1995) Chem Biol 2:325-333
80. Cuenoud B, Szostak JW (1995) Nature 375:611-614
81. Ekland EH, Szostak JW, Bartel DP (1995) Science 269:364-370
82. Hager AJ, Szostak JW (1997) Chem Biol 4:607-617
83. Wright MC, Joyce GF (1997) Science 276:614-617
84. Huang F, Yarus M (1997) Biochemistry 36:6557-6563
85. Huang F, Yarus M (1997) Proc Natl Acad Sci USA 94:8965-8969
86. Jenne A, Famulok M (1998) Chem Biol 5:23-34
87. Unrau PJ, Bartel DP (1998) Nature 395:260-263
88. Haldane JBS (1930) Enzymes Longmans, Green & Co., MIT Press
89. Pauling L (1946) Chem Eng News 24:1375

90. Jencks WP (1969) Catalysis in Chemistry and Enzymology McGraw Hill New York
91. Tramontano A, Janda KD, Lerner RA (1986) Science 234:1566–1570; Jacobs J, Schultz PG (1987) J Am Chem Soc 109:2174–2176; Schultz PG (1989) Angew. Chem. Int. Ed. 28: 1283–1295; Schultz PG, Lerner RA (1993) Acc Chem Res 26:391–395; Hilvert D (1994) Curr. Opin. Struct. Biol. 4: 612–617; Posner B, Smiley J, Lee I, Benkovic S (1994) Trends Biochem Sci 19:145–150; Jacobsen JR, Schultz PG (1995) Curr Opin Struct Biol 5: 818–824; Hsieh-Wilson LC, Xiang X-D, Schultz PG (1996) Acc Chem Res 29:164–170; Benkovic S (1996) Nature 383:23–24
92. Conn MM, Prudent JR, Schultz PG (1996) J Am Chem Soc 118:7012–7013
93. Li Y, Sen D (1996) Nature Struct Biol 3:743–747
94. Morris KN, Tarasow TM, Julin CM, Simons SL, Hilvert D, Gold L (1994) Proc Natl Acad Sci USA 91:13028–13032
95. Prudent JR, Schultz PG (1996) RNA catalysis and transition state stabilization, p 383–395. In: Eckstein F, Lilley DMJ (ed) Catalytic RNA, vol 10. Springer, Berlin Heidelberg New York
96. Li WX, Kaplan AV, Grant GW, Toole JJ, Leung LL (1994) Blood 83:677–682
97. Li Y, Geyer CR, Sen D (1996) Biochemistry 35:6911–6922
98. Li Y, Sen D (1997) Biochemistry 36:5589–5599
99. Li Y, Sen D (1998) Chem Biol 5:1–12
100. Wilson C, Szostak JW (1998) Chem Biol 5:609–617
101. Lehman N, Joyce GF (1993) Nature 361:182–185
102. Lorsch JR, Szostak JW (1994) Nature 371:31–36
103. Wilson C, Szostak JW (1995) Nature 374:777–782
104. Williams KP, Ciafré S, Tocchini-Valentini GP (1995) EMBO J 14:4551–4557
105. Lorsch J, Szostak JW (1995) Biochemistry 34:15315–15327
106. Illangasekare M, Sanchez G, Nickles T, Yarus M (1995) Science 267:643–647
107. Wecker M, Smith D, Gold L (1996) RNA 2:982–994
108. Frauendorf C, Jäschke A (1988) Angew Chem Int Ed 37:378–381
109. Beaudry AA, Joyce GF (1992) Science 257:635–641
110. Pan T, Uhlenbeck OC (1992) Biochemistry 31:3887–3895
111. Breaker RR, Joyce GF (1994) Chem Biol 1:223–229
112. Breaker RR, Joyce GF (1995) Chem Biol 2:655–660
113. Roth A, Breaker RR (1998) Proc Natl Acad Sci USA 95:6027–6031
114. Faulhammer D, Famulok M (1996) Angew Chem Int Ed 35:2837–2841
115. Jayasena VK, Gold L (1997) Proc Natl Acad Sci USA 94:10612–10617
116. Santoro SW, Joyce GF (1997) Proc Natl Acad Sci USA 94:4262–4266
117. Carmi N, Shultz LA, Breaker RR (1996) Chem Biol 3:1039–1046
118. Carmi N, Balkhi SR, Breaker RR (1998) Proc Natl Acad Sci USA 95:2233–2237
119. Ekland EH, Bartel DP (1996) Nature 382:373–376
120. Huang F, Yarus M (1997) Biochemistry 36:14107–14119
121. Huang F, Yang Z, Yarus M (1998) Chem Biol 5:669–678
122. Burmeister J, von Kiedrowski G, Ellington AD (1997) Angew Chem Int Ed 36:1321–1324
123. Illangasekare M, Yarus M (1997) J Mol Biol 268:631–639
124. Suga H, Szostak JW (1998) J Am Chem Soc 120:1151–1156
125. Ekland EH, Bartel DP (1995) Nucleic Acids Res 23:3231–3238
126. Spiegelman S (1967) Am Sci 55:221–264
127. Hager AJ, Pollard JD, Szostak JW (1996) Chem Biol 3:717–725
128. Connell GJ, Illangesekare M, Yarus M (1993) Biochemistry 32:5497–5502
129. Famulok M (1994) J Am Chem Soc 116:1698–1706
130. Majerfeld I, Yarus M (1994) Nat Struct Biol 1:287–292
131. Yang Y, Kochoyan M, Burgstaller P, Westhof E, Famulok M (1996) Science 272:1343–1347
132. Majerfeld I, Yarus M (1998) RNA 4:471–478
133. Illangasekare M, Yarus M (1997) J Mol Biol 274:519–529
134. Eaton BE (1997) Curr Opin Chem Biol 1:10–16
135. Service RF (1998) Science 282:1020–1021

136. Sakthivel K, Barbas III CF (1998) Angew Chem Int Ed 37:2872–2875
137. Otto S, Bertoncin F, Engberts JBFN (1996) J Am Chem Soc 118:7702–7707
138. Igloi GL (1988) Biochemistry 27:3842–3849
139. Sassanfar M, Szostak JW (1993) Nature 364:550–553
140. Jiang F, Kumar RA, Jones RA, Patel DJ (1996) Nature 382:183–186
141. Ellington AD, Szostak JW (1992) Nature 355:850–852
142. Burgstaller P, Famulok M (1995) Angew Chem Int Ed 34:1189–1192
143. Burgstaller P, Famulok M (1998) Synthetic ribozymes and deoxyribozymes, in press.
     In: Waldmann H, Mulzer J (ed) Organic Synthesis Highlights, vol 3. Wiley-VCH, Weinheim
144. Michel F, Hanna M, Green R, Bartel DP, Szostak JW (1989) Nature 342:391–395
145. Joyce GF (1998) Proc Natl Acad Sci USA 95:5845–5847
146. Patel DJ (1997) Curr Opin Chem Biol 1:32–46
147. Patel DJ, Suri AK, Jiang F, Liang L, Fan P, Kumar RA, Nonin S (1997) J Mol Biol 272:
     645–664
148. Burgstaller P, Famulok M (1994) Angew Chem Int Ed 33:1084–1087
149. Burgstaller P, Famulok M (1996) Bioorg Med Chem Lett 6:1157–1162
150. Fan P, Suri AK, Fiala R, Live D, Patel DJ (1996) J Mol Biol 258:480–500
151. Tang J, Breaker RR (1998) Nucleic Acids Res 26:4214–4221
152. Joyce GF, Orgel LE (1993) Prospects for the understanding of the RNA world, p 1–25.
     In: Gesteland RF, Atkins JF (ed) The RNA World. Cold Spring Harbor Laboratory Press,
     Cold Spring Harbor, New York

Author Index Volume 201–202

The volume numbers are printed in italics